Strigolactones - Biology and Applications

Hinanit Koltai • Cristina Prandi
Editors

Strigolactones - Biology and Applications

 Springer

Editors
Hinanit Koltai
Institute of Plant Science
Agricultural Research Organization (ARO)
Volcani Center
Rishon LeZion, Israel

Cristina Prandi
Department of Chemistry
Università degli Studi di Torino
Torino, Italy

ISBN 978-3-030-12155-6   ISBN 978-3-030-12153-2   (eBook)
https://doi.org/10.1007/978-3-030-12153-2

Library of Congress Control Number: 2019934446

This Springer imprint is published by the registered company Springer Nature Switzerland AG.
The registered company address is: Gewerbestrasse 11, 6330 Cham, Switzerland

Acknowledgements

We strongly believe that this supplementary textbook, both printed book and e-book, may be attractive for graduate and postgraduate students and may help teachers and lecturers to better present and summarize strigolactone-related science and the new emerging concepts.

Finally, we would like to thank all the authors and reviewers for accepting our invitation to contribute to this textbook, the first published so far and entirely dedicated to SLs. Special thanks go to all those scientists who are not formally authors of this book but with their research and their outstanding publications contributed and are still contributing to dissect the still unveiled aspects of this challenging class of molecules and paved the way to possible exploitation in many different fields.

We had the opportunity and pleasure to personally meet most of these scientists in the framework of the COST Action FA1206 "Strigolactones: biological roles and applications". The numerous meetings, workshops and Short-Term Scientific Missions organized thanks to the COST tool allowed us to create synergies among scientists of different expertise and cultural background, enabled new collaborations and established a community of people working on common and shared aims in the name of science.

Introduction

Strigolactones: New Plant Hormones and Much More. . .

Plants produce and release various chemicals into the environment, as well as primary and secondary metabolites. Abiotic and biotic stresses affect the composition and the amount of these compounds by promoting or suppressing their biosynthesis and/or efflux.

Strigolactones (SLs) are typical examples of such signalling molecules. Plants release only very small amounts of SLs into the soil, and these molecules decompose rapidly in the rhizosphere. SLs can only be analysed and quantified using recently developed highly sensitive mass spectrometry methods and were originally isolated as germination stimulants for seeds of parasitic weeds of the family *Orobanchaceae*. Therefore, these compounds were regarded as harmful secondary metabolites since they were detrimental to the producing plant. It has been subsequently shown that SLs act as indispensable chemical signals for root colonization by symbiotic arbuscular mycorrhizal (AM) fungi and then became recognized as beneficial plant metabolites.

However, only recently they were recognized as plant hormones that regulate different aspects of plant development, among others as mediators of plant response to abiotic conditions. This recognition led to a dramatic increase in the interest in these new plant hormones, and to thriving research on different biological aspects of these hormones, from different disciplinary fields including their signal transduction, reception and biosynthesis, evolution and genetic regulation. This blooming research unveiled both already existing and new biological concepts, such as redefinition of plant hormones and their crosstalk, new functional diversity of receptors, evolvement of plants to parasitic life habit, smoke and hormone mirrors, core signalling pathways and even phloem transport of receptor protein. Another important aspect of SLs is their developed synthetic chemistry and the opening of a variety of potential applications in agriculture and medicine.

Yet, despite the thriving scientific activity on SLs, our and other scientists' experience suggested that many university and college students and lecturers of plant sciences are not fully aware of SL-related sciences. As a result, these subjects are not being properly conveyed to the next-generation scientists in many different countries.

The above considerations led to the idea of composing a supplementary graduate textbook that addresses teachers, lecturers and biology and agronomy students. The challenge in undertaking the project of writing a textbook on SLs was to write a book whose structure develops around ideas from very different disciplinary fields, rather than presenting a sequence of facts.

We decided to organize the book into six chapters, from SL biosynthesis and perception, to the role of SLs as plant hormones, as parasitic weed germination and hyphal branching inducers respectively, to the chemistry and the stereochemical aspects of natural and synthetic SLs. A full chapter has been dedicated to the involvement of SLs in evolution aspects. Each chapter has been conceived to stand by itself with its general introduction, which enables the reader to look deeply into the specific aspects addressed by every single chapter. Each chapter conveys a certain topic to allow a broad view on each of the presented subjects. Authors and co-authors are the leading scientists in these subjects from around the world and were able to give an accurate, deep and comprehensive view of the subjects. Glossary and synopses that may ease comprehension of the related terms and concepts are also included in each chapter. Given that SLs are a cutting-edge topic nowadays and the literature updates every day, the most relevant literature references embedded in the text enable even non-expert readers to go directly to the focus of their interest. Illustration and figures are provided to better demonstrate the presented topics.

Chapter 1: Strigolactone Biosynthesis and Signal Transduction

Kun-Peng Jia, Changsheng Li, Harro J. Bouwmeester, and Salim Al-Babili

In this chapter, the authors provide an overview on the enormous progress that has been recently made in elucidating SL biosynthesis and signal transduction. They described the tailoring pathway from the carotenoid precursor to the central intermediate carlactone, highlighting the stereo-specificity of the involved enzymes, the all-*trans*/9-*cis*-β-carotene isomerase (D27), the 9-*cis*-specific CAROTENOID CLEAVAGE DIOXYGENASE 7 (CCD7) as well as CCD8 and its unusual catalytic activity. They then outline the oxidation of carlactone by cytochrome P450 enzymes, such as the Arabidopsis MORE AXILLARY GROWTH 1 (MAX1), into different SLs and the role of other enzymes in generating this diversity, and discuss why plants produce many different SLs. This is followed by depicting hormonal and nutritional factors that regulate SL biosynthesis and release and by a description of transport mechanisms. In the second part of the chapter, the authors focus on SL perception and signal transduction, describing the SL receptor DECREASED APICAL DOMINANCE 2 (DAD2)/DWARF14 (D14) and its unique features, the central function of protein degradation mediated by the F-box protein MAX2 and its homologues. They also discuss the latest advances in understanding how SLs regulate the transcription of target genes and the role of SMXL/D53 transcription inhibitors.

Chapter 2: Strigolactones as Plant Hormones

Catherine Rameau, Sofie Goormachtig, Francesca Cardinale, Tom Bennett, and Pilar Cubas

This chapter presents SL activity as a new class of plant hormones. The authors present evidence to support a role for SLs in regulating aerial and underground plant architecture: they repress shoot branching, promote internode elongation and height, affect gravitropic setpoint angle, control secondary growth in stems and affect leaf shape and leaf serration, reproductive organ size, control of flowering time, leaf morphology and tuberization. SLs also regulate root architecture, adventitious root development and root hair development. SLs are involved in plant response to abiotic stress, including nutrient deprivation and osmotic stress. Also, SL crosstalk with other plant hormones is introduced and discussed. An emphasis is being put on "direct" interactions between SLs and other hormones, and this is rationalized in terms of SL functionality. SLs and interacting hormones are characterized as a systemically acting platform that regulates development and responses to soil conditions. It is stated that the impact of SLs on key developmental processes, such as plant architecture and their involvement in the acclimation of plants to environmental stresses, raises the possibility of using these hormones and signalling pathways as agricultural tools to optimize crop plant architecture and resilience to abiotic stress.

Chapter 3: Strigolactones and Parasitic Plants

Maurizio Vurro, Angela Boari, Benjamin Thiombiano, and Harro Bouwmeester

A parasitic plant is a flowering plant that attaches itself morphologically and physiologically to a host (another plant) by a modified root (the haustorium). Only about 25 out of the 270 genera of parasitic plants have a negative impact on agriculture and forestry and thus can be considered weeds. Among them, the most damaging root-parasitic weeds belong to the genera *Orobanche* and *Phelipanche* (commonly named broomrapes) and *Striga* (witchweeds) (all belonging to the Orobanchaceae family). Considering the aim of the book, this chapter focuses only on this group of parasitic weeds, as in these plants SLs have a key role both in their life cycle and in management strategies to control them. Distribution, agricultural importance and life cycle of these parasitic weeds are briefly introduced, after which the authors focused on the role of SLs in seed germination, parasite development, host specificity, plant nutrition and microbiome composition. Furthermore, some weed control approaches involving SLs are discussed.

Chapter 4: The Role of Strigolactones in Plant–Microbe Interactions

Soizic Rochange, Sofie Goormachtig, Juan Antonio Lopez-Raez, and Caroline Gutjahr

Plants associate with an infinite number of microorganisms that interact with their hosts in a mutualistic or parasitic manner. Evidence is accumulating that SLs play a role in shaping these associations. The best described function of SLs in plant–microbe interactions is in the rhizosphere, where, after being exuded from the root, they activate hyphal branching, enhanced growth and energy metabolism of symbiotic arbuscular mycorrhizal fungi (AMF). Furthermore, an impact of SLs on the quantitative development of root nodule symbiosis with symbiotic nitrogen-fixing bacteria and on the success of fungal and bacterial leaf pathogens is beginning to be revealed. Thus far, the role of SLs has predominantly been studied in binary plant–microbe interactions. It can be predicted that their impact on the bacterial, fungal and oomycetal communities (microbiomes), which thrive on roots, in the rhizosphere and on aerial tissues, will be addressed in the near future.

Chapter 5: Evolution of Strigolactone Biosynthesis and Signalling

Sandrine Bonhomme and Mark Waters

In this chapter, the authors present the current knowledge on when and how SLs originated and what functions they have in non-seed plants. Although this field is still much in its infancy, rapid advances are being made in the acquisition and interpretation of data and information. These advances lead to several emerging concepts that are conveyed in this chapter. The evolution of land plants is suggested to be associated with increases in developmental complexity, brought about by diversification of gene families and hormone signalling pathways. Good model species for the study of early diverging land plants are the moss *Physcomitrella patens* and the liverwort *Marchantia polymorpha*. Also, with minor exceptions, the core enzymes for SL biosynthesis via carlactone are present in all land plants. This suggests that SLs, or SL-like compounds, are common to all land plants. However, while SLs have a clear developmental/hormonal role in angiosperms, the function of SLs in early land plants is equivocal. Some receptor enzymes that perceive SLs and/or SL-like compounds predate land plants and have undergone substantial duplication throughout land plant evolution, whereas others essential for SL perception in angiosperms are not required in moss. It is also presented and discussed that parasitic weeds demonstrate evolution in action.

Chapter 6: The Chemistry of Strigolactones

Cristina Prandi and Christopher S. P. McErlean

SLs are a group of small molecules which were first reported after isolation from the root exudates of cotton in 1966. These compounds were potent germination stimulants of the parasitic witchweed (*Striga lutea* Lour.), which has an economically devastating effect on many important crops. As such, SLs captured the attention of chemists who sought to (1) determine the structures of these molecules, (2) synthesize the molecules and (3) synthesize molecules that mimic the biological actions of natural SLs. This chapter highlights the progress that has been made in each of these areas, which can collectively be categorized as "the chemistry of SLs".

Contents

About the Editors

Prof. Hinanit Koltai, PhD, is a Senior Research Scientist at the Agricultural Research Organization, Volcani Center, Israel. She is the Editor of books and a member of Editorial boards in international scientific journals. She is a leading Author of more than 80 peer reviewed publications and more than 30 book chapters and invited reviews and she holds more than 10 patents. Research in Koltai lab is focused on plant hormones and medicinal plants, specifying their medical influence at the molecular level on human cells and tissue. She teaches plant development and medical cannabis courses in Bar Ilan University, Israel.

Cristina Prandi is a Full Professor of Organic Chemistry at the University of Torino, where she also serves as Deputy Director for Research. Her main interests are in organometallic chemistry, gold catalysis and target-oriented synthesis. She has conducted research on the synthesis of bioactive phytohormone analogues, focusing on SAR (structure–activity relationship) studies and the design of active derivatives. Recently, she has also investigated the use of plant metabolite analogues for their potential anticancer benefits. She is the author of more than 100 scientific publications and holds three patents.

Prof. C. Prandi and Prof. H. Koltai chaired a COST ACTION exclusively dedicated to strigolactones (FA1206 2012–2017, Strigolactones, biological roles and applications).

Chapter 1
Strigolactone Biosynthesis and Signal Transduction

Kun-Peng Jia, Changsheng Li, Harro J. Bouwmeester, and Salim Al-Babili

Abstract Strigolactones (SLs) are a group of carotenoid derivatives that act as a hormone regulating plant development and response to environmental stimuli. SLs are also released into soil as a signal indicating the presence of a host for symbiotic arbuscular mycorrhizal fungi and root parasitic weeds. In this chapter, we provide an overview on the enormous progress that has been recently made in elucidating SL biosynthesis and signal transduction. We describe the tailoring pathway from the carotenoid precursor to the central intermediate carlactone, highlighting the stereo-specificity of the involved enzymes, the all-*trans*/9-*cis*-β-carotene isomerase (D27), the 9-*cis*-specific CAROTENOID CLEAVAGE DIOXYGENASE 7 (CCD7), as well as CCD8 and its unusual catalytic activity. We then outline the oxidation of carlactone by cytochrome P450 enzymes, such as the Arabidopsis MORE AXIL-LARY GROWTH 1 (MAX1), into different SLs and the role of other enzymes in generating this diversity, and discuss why plants produce many different SLs. This is followed by depicting hormonal and nutritional factors that regulate SL biosynthesis and release, and by a description of transport mechanisms. In the second part of our chapter, we focus on SL perception and signal transduction, describing the SL receptor DECREASED APICAL DOMINANCE 2 (DAD2)/DWARF14 (D14) and its unique features, the central function of protein degradation mediated by the F-box protein MAX2 and its homologs. We also discuss the latest advances in understanding how SLs regulate the transcription of target genes and the role of SMXL/D53 transcription inhibitors.

Keywords Strigolactone biosynthesis · Carotenoids · D27 · CCD7 · CCD8 · Carlactone · MAX1 · Strigolactone signaling · Strigolactone perception

K.-P. Jia · S. Al-Babili (✉)
The BioActives Lab, The King Abdullah University of Science and Technology (KAUST), Thuwal, Saudi Arabia
e-mail: kupeng.jia@kaust.edu.sa; salim.babili@kaust.edu.sa

C. Li · H. J. Bouwmeester
Plant Hormone Biology Group, Swammerdam Institute for Life Sciences, University of Amsterdam, Amsterdam, The Netherlands
e-mail: c.li3@uva.nl; H.J.Bouwmeester@uva.nl

© Springer Nature Switzerland AG 2019
H. Koltai, C. Prandi (eds.), *Strigolactones - Biology and Applications*,
https://doi.org/10.1007/978-3-030-12153-2_1

1

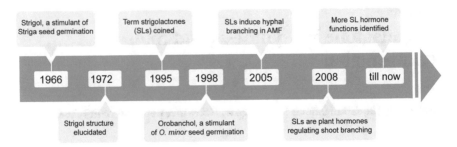

Fig. 1.1 Timeline of discovery of SLs and their biological functions. *AMF* arbuscular mycorrhizal fungi

1.1 Introduction

1.1.1 Strigolactones: An Overview

Plants use chemical signals, such as hormones, to coordinate growth and developmental processes as well as responses to environmental stimuli. In addition, chemical signals are the major means in the communication of plants with other organisms in their environment. Strigolactones (SLs) are an exceptional example because of their dual activity as plant hormone with various developmental and stress-related functions and as communication signal in the rhizosphere (Akiyama et al. 2005; Cook et al. 1966; Gomez-Roldan et al. 2008; Siame et al. 1993; Umehara et al. 2008). Interestingly, it was the latter activity that led to the discovery of SLs. Strigol, the first identified SL, was isolated as a $C_{19}H_{22}O_6$ compound in 1966 from root exudates of cotton as seed germination stimulant of the root parasitic weed *Striga lutea* (Cook et al. 1966), but its structure was only elucidated 6 years later (Cook et al. 1972). Since then, a series of structurally related compounds have been isolated from root exudates of different plant species, based on their capability to trigger seed germination in root parasitic plants, i.e., *Striga*, *Orobanche*, *Alectra*, and *Phelipanche* spp. (Yoneyama et al. 2013; Xie et al. 2010). The collective name strigolactone was coined almost 30 years after the discovery of strigol to designate this group of intriguing compounds (Butler 1995). Figure 1.1 briefly introduces a timeline of discovery of SLs and of their various biological functions.

Root parasitic plants are obligate parasites that have partially or completely lost their photosynthetic capacity during the evolution toward parasitism (Bouwmeester et al. 2003). These weeds produce enormous numbers of tiny seeds containing few reserves that can ensure survival of the seedlings after germination for only few days. Hence, seed germination in *Striga* and related root parasitic species is tightly regulated and takes place only if a host root is present in close vicinity. This synchronization is brought about by the strict germination dependency on host released chemical signals, in most cases SLs. Following germination, seedlings develop a haustorium to connect to the host root and siphon off water and minerals and assimilate (Bouwmeester et al. 2003; Delavault et al. 2002; Xie et al. 2010). This parasitism stunts the growth of the host. Thus, infestation of crops by root parasitic plants is a severe problem for agriculture in warm and temperate zones of Europe, Asia, and Africa, causing enormous yield losses in crops, such as cereals, legumes, rapeseed, tomato, and sunflower (Parker 2009).

The question why plants release SLs into the rhizosphere, though they provoke thereby the attack of parasitic weeds, had been bothering plant scientists for decades. In 2005, Akiyama et al. provided the answer to this problem by showing that SLs are a crucial factor in establishing the symbiosis of roots with beneficial arbuscular mycorrhizal (AM) fungi (Akiyama et al. 2005). SLs induce branching of hyphae (threadlike filaments) in AM fungi, allowing them to grow toward and colonize the host root. AM symbiosis is an important support for the growth and survival of land plants, as evidenced by its presence in around 80% of them (Bonfante and Genre 2015; Gutjahr and Parniske 2013). This symbiosis plays a key role in supplying the host plant with minerals absorbed by the fungal hyphae that extend the plant's roots system, enabling the exploitation of a much larger soil volume. In return, the heterotrophic fungal partner obtains reduced carbon in the form of sugars that are produced by photosynthesis and which are used as energy and carbon source (Bonfante and Genre 2015; Gutjahr and Parniske 2013). The tremendous benefits of AM symbiosis in natural ecosystems explain why plants, particularly under phosphorus deficiency, release SLs into soil (Gutjahr 2014; Khosla and Nelson 2016). It can be assumed that root parasitic weeds, which evolved much later than the AM symbiosis, have coopted the SL signal as reliable indicator for host presence, by evolving a highly sensitive SL detection system coupled with the induction of seed germination.

The function of SLs as plant hormone was reported 3 years after revealing their role in AM symbiosis (Gomez-Roldan et al. 2008; Umehara et al. 2008). This discovery became possible due to the availability of increased shoot branching and high-tillering mutants from several plant species. Genetic analysis and grafting studies enabled the classification of these mutants into two groups. The first one is deficient in the synthesis of a postulated mobile shoot branching inhibitory signal, while the second one is affected in the perception of this signal. The lack of SLs in mutants from the first group and the capability of the SL analog GR24 to rescue their high-branching/high-tillering phenotype suggested that SLs are the long sought-after shoot branching inhibitory signal. This conclusion was further confirmed by the lack of response to GR24 in the supposed perception mutants (Gomez-Roldan et al. 2008; Umehara et al. 2008). Nowadays, SLs are recognized as a plant hormone that mediates the adaptation to nutrient deficiency and is involved (beyond attracting AM fungi as helpers in nutrient acquisition) in different aspects of plant growth and development, such as root development, stem secondary growth, and senescence (Al-Babili and Bouwmeester 2015; Brewer et al. 2013; Jia et al. 2018; Koltai 2011; Ruyter-Spira et al. 2013; Waters et al. 2017). In addition, several lines of evidence suggest a role of SLs in plant response to biotic and abiotic stresses (Decker et al. 2017; Liu et al. 2015; Torres-Vera et al. 2014; Van Ha et al. 2014) (Fig. 1.2 illustrates the major biological functions of SLs).

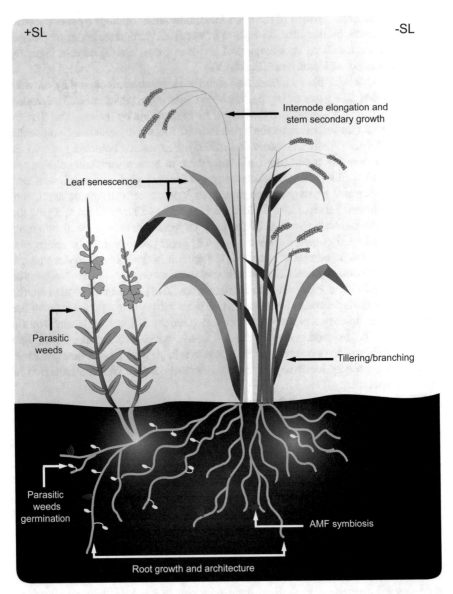

Fig. 1.2 Biological functions of SLs. SLs inhibit the outgrowth of axillary buds, which prevents the formation of tillers/branches. SLs also accelerate leaf senescence, promote the growth of internode elongation and stem secondary growth, and determine root growth and architecture. Released into the rhizosphere, SLs induce seed germination of root parasitic weeds, such as *Striga* spp. and induce hyphal branching in symbiotic arbuscular mycorrhizal fungi (AMF)

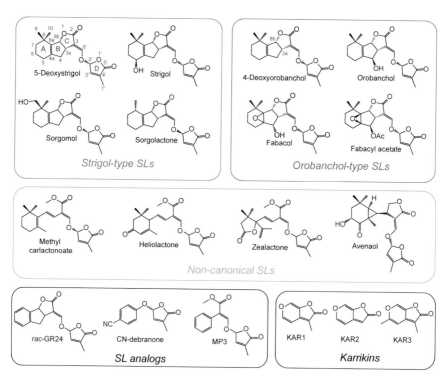

Fig. 1.3 Structures and classification of SLs and analogs. Canonical SLs are characterized by the presence of a tricyclic lactone (ABC-ring) connected to a second lactone (D-ring) in *R*-configuration ($2'$ *R*). They are divided into strigol- and orobanchol-like SLs based on the stereochemistry of the B-/C-ring junction, with the C-ring in β orientation (up) and α orientation (down) in strigol-like and orobanchol-like SLs, respectively. Non-canonical SLs contain different structural elements instead of the tricyclic lactone. *rac*-GR24 is the most commonly used SL analog; CN-debranone and methyl phenlactonoate 3 (MP3) are two SL analogs with a simple structure; KAR1, KAR2, and KAR3 are examples of karrikins

1.1.2   SL Examples, Structures, and Nomenclature

So far, there are around 25 characterized natural SLs, and it can be expected that this number will increase in the future (Xie 2016). The structure of strigol as the first-discovered SL was also the first-elucidated (Cook et al. 1972, Fig. 1.3). This compound was also later identified as the major *Striga* seed germination stimulant in maize root exudates (Siame et al. 1993) and shown to be present in moonseed root culture filtrate (Yasuda et al. 2003). Interestingly, maize also produces 5-deoxystrigol, sorgomol (Fig. 1.3), and, as recently shown, two novel SLs that were designated as zealactone and zeapyranolactone (see Sect. 1.2.5) (Yoneyama et al. 2015; Charnikhova et al. 2017; Xie et al. 2017). Similarly, sorghum produces and exudes various structurally different SLs, including strigol, sorgolactone, 5-deoxystrigol, and sorgomol that were identified in root exudates (Siame et al.

1993; Awad et al. 2006; Xie et al. 2008; Jamil et al. 2013; Motonami et al. 2013). Heliolactone (Fig. 1.3) is a further example for a natural SL, which was isolated as seed germination stimulant for both *Orobanche cumana* and *Striga hermonthica* from root exudates of sunflower (Ueno et al. 2014). Orobanchol (Fig. 1.3), the first discovered seed germination stimulant for *Orobanche minor*, was isolated together with its acetate, alectrol, from red clover (Yokota et al. 1998). Fabacyl acetate, identified in pea root exudates, is the first identified SL containing an epoxide group (Xie et al. 2009). Solanacol, which was identified in tobacco and tomato, is the first natural strigolactone equipped with a benzene ring (Xie et al. 2007).

These and other natural SLs are defined as carotenoid derivatives characterized by a structure consisting of a conserved butenolide ring (D-ring, Fig. 1.3) that is linked in a defined stereochemical configuration (*R*-configuration) to a second, variable moiety (Jia et al. 2018). Based on the structure of the second part, SLs are considered as canonical if they contain a conserved tricyclic lactone (ABC-ring, Fig. 1.3) or non-canonical if they have a different structure instead. Canonical SLs are divided based on the stereochemistry of the BC-ring junction into the orobanchol-type SLs with the C-ring in α-orientation (8b*R*-configuration) and the strigol-type SLs, with the C-ring in β-orientation (up; 8b*S*-configuration) (Jia et al. 2018, Fig. 1.3). Strigol and orobanchol were named according to their activity in triggering seed germination in *Striga* and *Orobanche* species and derive from the corresponding precursors 5-deoxystrigol and ent-2′-*epi*-5-deoxystrigol (4-deoxyorobanchol, see Sect. 1.2.5), respectively (Cook et al. 1972; Siame et al. 1993; Yokota et al. 1998). Strigol and orobanchol act as references to designate other structurally related SLs that differ by substitution (s) or in the stereo-configuration of chiral center at the C2′ atom and/or the B/C junction. The stereo-configuration is usually described by the abbreviations *ent-* and *epi-* referring to enantiomer (a mirror image of the reference) and epimer (opposite orientation at a single C atom), respectively. However, the usage of these abbreviations may in some cases be confusing. Therefore, there is a tendency to replace complicated names by simple ones, e.g., 4-deoxyorobanchol instead of *ent-2'-epi*-5-deoxystrigol. Non-canonical SLs are distinguished by the presence of different structures as a second moiety instead of the ABC-ring characteristic for all canonical SLs (Al-Babili and Bouwmeester 2015; Jia et al. 2018). Methyl carlactonoate (Abe et al. 2014), heliolactone (Ueno et al. 2014), zealactone and zeapyranolactone (Charnikhova et al. 2017, 2018; Xie et al. 2017), and avenaol (Kim et al. 2014) are examples for this type of SLs (Fig. 1.3). The variability of non-canonical SLs and modifications of the A- and B-ring in canonical SLs, which include hydroxylation, methylation, epoxidation, and ketolation, give rise to the diversity of natural SLs (Jia et al. 2018; Wang and Bouwmeester 2018).

Synthesis of natural SLs is laborious because of their complex structure and the presence of chiral centers. In addition, plants release SLs in minute concentrations, which make the isolation of large quantities of these compounds from natural sources almost impossible. Therefore, SL research and application have been depending on synthetic analogs, such as the widely used GR24 (Fig. 1.3) that contains an ABC-ring (with a phenolic A-ring) coupled to the characteristic D-ring. However, in the meanwhile there are more simple analogs with considerable biological activity. CN-debranone (Fukui et al. 2011), nitro-phenlactone (Jia et al. 2016), and methyl phenlactonoates (MP3, Jamil et al. 2017) are examples for such compounds (Fig. 1.3).

There is a different class of signaling molecules, the karrikins that share the butenolide ring as a conserved structural element with SLs (Fig. 1.3). They are perceived by a homologous signal transduction pathway with some common signal transduction components (Waters et al. 2013). Karrikins were first isolated from the smoke of burning plant material as compounds triggering seed germination in pioneer plants that emerge after forest and bush fires (Nelson et al. 2009). However, karrikins do not induce seed germination in root parasitic plants, in which the karrikin receptor gene has been presumably subjected to repeated duplication during the evolution of parasitism, accompanied by changing the ligand specificity toward SLs (Conn et al. 2015; Tsuchiya et al. 2015). Karrikins are supposed to mimic a yet unidentified signaling molecule that regulates developmental processes, such as seed dormancy and photomorphogenesis different from those governed by SLs (Flematti et al. 2015; Nelson et al. 2010). It should be mentioned here that the common SL analog GR24 is usually available as a racemic mixture of the two enantiomers GR24 and *ent*-GR24, which bind to the SL receptor D14 and the karrikin receptor D14-like 1 (KAI2), respectively (Scaffidi et al. 2014). Therefore, *rac*-GR24 activates both SL and karrikin signaling pathways.

1.2 SL Biosynthesis

1.2.1 Sesquiterpenes or Carotenoid Derivatives?

For almost 40 years after the discovery of the first SL, there was no progress in elucidating SL biosynthesis because of the uncertainty regarding the metabolic origin of these compounds. SLs were supposed to be sesquiterpene lactones (Fig. 1.4) (Butler 1995; Yokota et al. 1998), a class of secondary metabolites with medical importance, which originate from the cytosolic C_{15} compound farnesyl diphosphate (de Kraker et al. 1998). However, the structural similarity of the A-ring of SLs with the ionone rings in carotenoids led to the hypothesis that SLs may originate from carotenoids, like the plant hormone abscisic acid (ABA) (Fig. 1.4, Bouwmeester et al. 2003; Parry and Horgan 1992; Tan et al. 1997). Both carotenoids and sesquiterpenes are isoprenoids, i.e., they derive from the universal C_5 building block isopentenyl diphosphate (IPP) (Fig. 1.4). However, plants utilize two different pathways for IPP synthesis: the mevalonate pathway that leads to sesqui- (C_{15}) and triterpenes (C_{30}), steroids and other cytosolic isoprenoids, and the plastid 2-C-methyl-D-erythritol 4-phosphate (MEP) pathway that provides the IPP precursor for the biosynthesis of plastid isoprenoids, such as mono- and diterpenes (C_{20}), chlorophylls, carotenoids, and tocopherols. To answer the question on the SL metabolic origin, Matusova et al. 2005 applied mevastatin and fosmidomycin to maize seedlings, which are isoprenoid biosynthesis inhibitors

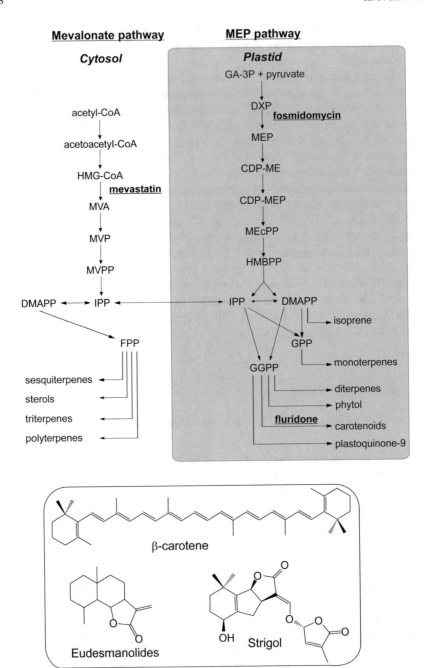

Fig. 1.4 Plant isoprenoid biosynthesis pathways (adapted from Matusova et al. 2005). Plants utilize two different IPP and DMAPP biosynthesis routes: the mevalonate pathway and the MEP pathway (also called as non-mevalonate pathway). The two pathways are present in different compartments and are used to produce the IPP building block for diverse classes of isoprenoids, such as carotenoids and sesquiterpene lactones. Structures of examples of these two classes are shown in the inset next to the structure of strigol. Abbreviations: HMG-CoA, 3-hydroxy-3-methylglutaryl CoA; MVA,

specific for the mevalonate and MEP pathway, respectively (Fig. 1.4, Matusova et al. 2005). In addition, they used fluridone, an inhibitor of carotenoid biosynthesis (Fig. 1.4). After treatment, Matusova et al. 2005 investigated the *Striga hermonthica* seed germination stimulating activity of root exudates collected from treated seedlings, as proxy for the content of SLs. Intriguingly, the *Striga* seeds germination stimulating activity was greatly decreased in root exudates of fluridone-treated maize (Matusova et al. 2005). Similar results were also obtained for fluridone-treated cowpea and sorghum root exudates that induced much lower germination of seeds of *Orobanche crenata* and *Striga hermonthica,* respectively. Because fluridone inhibits the enzymatic activity of phytoene desaturase, which catalyzes an early step in carotenoid biosynthesis, root exudates of several maize carotenoid biosynthesis-deficient mutants (*lw1, y10, al1y3, vp5, y9,* and *vp14-2274*) were tested for their parasitic plant seed germination stimulating activity. Indeed, root exudates of all these mutants induced lower germination of *Striga* seeds than those of wild-type seedlings (Matusova et al. 2005). Taken together, these results indicated that SLs likely derive from carotenoids or apocarotenoids (carotenoid cleavage products) rather than sesquiterpene lactones.

1.2.2   Carotenoids: An Overview

The attractive colors of many fruits and flowers are the result of carotenoid accumulation. However, these widespread isoprenoid pigments also fulfill more important and vital functions in plants and other photosynthetic organisms (DellaPenna and Pogson 2006; Walter and Strack 2011). Carotenoids are indispensable constituents of the photosynthetic apparatus, protecting chlorophyll and other cellular components from photooxidation and contributing to the light-harvesting process (Hashimot et al. 2016). Carotenoids are split into two classes: the oxygen-free carotenes and the xanthophylls that carry oxygen-containing functional groups (Fig. 1.5). Besides this classification, carotenoids are distinguished by the type of their end groups (acyclic or linear, monocyclic, and bicyclic) and by the stereo-configuration of their double bonds. In plants, carotenoid biosynthesis takes place in plastids and is initiated by the condensation of two molecules of geranylgeranyl diphosphate (C_{20}) to 15-*cis*-phytoene (C_{40}), which is catalyzed by the key enzyme phytoene synthase. Phytoene contains only three conjugated double bonds and is, hence, colorless. Phytoene is

---

Fig. 1.4 (continued) mevalonic acid; MVP, mevalonate-5-phosphate; MVPP, mevalonate-5-pyrophosphate; FPP, farnesyl pyrophosphate; GA-3P, D-glyceraldehyde-3-phosphate; DXP, 1-deoxy-D-xylulose 5-phosphate; MEP, 2-C- methyl-D-erythritol 4-phosphate; CDP-MEP, CDP-ME 2-phosphate; MEcPP, 2-C-methyl-D-erythritol 2,4-cyclopyrophosphate; HMBPP, 1-hydroxy-2-methyl-2-butenyl 4-pyrophosphate; GPP, geranyl pyrophosphate; GGPP, geranylgeranyl pyrophosphate; IPP, isopentenyl pyrophosphate; DMAPP, dimethylallyl pyrophosphate

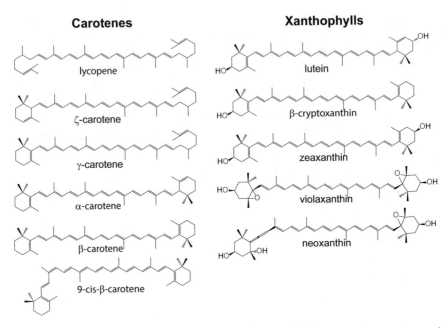

Fig. 1.5 Representative examples of the two classes of carotenoids: the oxygen-free carotenes and the xanthophylls that carry oxygen-containing functional groups

transformed in the next steps by a sequence of desaturation and *cis/trans* isomerization reactions into the red linear all-*trans*-lycopene that is equipped with 11 conjugated double bonds (Fig. 1.5). Cyclization reactions convert all-*trans*-lycopene into all-*trans*-β-carotene (Fig. 1.5), which carries two β-ionone rings, and all-*trans*-α-carotene (Fig. 1.5) equipped with an ε- and a β-ionone ring, dividing the pathway into the α- and β-branches. Hydroxylation of the two ionone rings in α- and β-carotene leads to all-*trans*-lutein and all-*trans*-zeaxanthin, respectively (Fig. 1.5). Epoxidation of the ionone rings in zeaxanthin forms violaxanthin (Fig. 1.5) via the mono-epoxidated antheraxanthin. All-*trans*-violaxanthin is the precursor of all-*trans*-neoxanthin (Fig. 1.5), the final product of the β-branch in plant carotenoid biosynthesis, which contains an allenic double bond (DellaPenna and Pogson 2006; Fraser and Bramley 2004; Moise et al. 2014; Walter and Strack 2011). The composition and amounts of carotenoids in different tissues are variable and depend on the type of plastids (Howitt and Pogson 2006; Ruiz-Sola and Rodríguez-Concepción 2012). Chromoplasts, which contain specialized structures to accommodate high amounts of carotenoids, accumulate certain types of carotenoids, such as lutein in daffodil flowers or lycopene in tomato fruits (DellaPenna and Pogson 2006; Fraser and Bramley 2004), while the photosynthetically active chloroplasts have more defined carotenoid composition consisting of about 45% lutein, 25–30% β-carotene, and 10–15% of each of violaxanthin and neoxanthin (Goodwin 1988; Lakshminarayana et al. 2005). Root leucoplasts contain only low amounts of

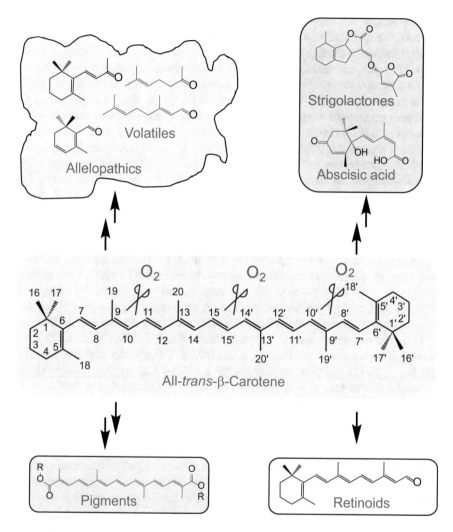

Fig. 1.6 Carotenoids are susceptible to oxidative cleavage that leads to a wide range of different compounds generally called apocarotenoids. The cleavage reaction is mediated by carotenoids cleavage dioxygenases (CCDs) but can also occur spontaneously triggered by reactive oxygen species. The structure of all-*trans*-β-carotene depicts the numbering of the C atoms, which is used to designate the cleavage site and the resulting apocarotenoid

carotenoids composed of around 30% lutein, 25% β-carotene, and 45% β-xanthophylls, as was shown for Arabidopsis (Britton 1995; Maass et al. 2009).

The conjugated double-bond system of carotenoids, which is responsible for their color and functions in photosynthesis, makes them prone to oxidation (Fig. 1.6). This process leads to cleavage of carotenoids into smaller molecules that carry carbonyl groups and which are generally called apocarotenoids (Giuliano et al. 2003; Nisar et al. 2015; Walter and Strack 2011). Carotenoid cleavage can occur at each double bond in the carotenoid backbone, triggered by reactive oxygen species (ROSs) that arise particularly under stress conditions. However, plants and other organisms, including humans, utilize an enzymatic cleavage of defined double bonds in specific carotenoids to generate particular apocarotenoids that exert important biological functions. Prominent examples of such apocarotenoids are the vision chromophore and vitamin A precursor retinal (Moise et al. 2005) and precursors for the fungal pheromone trisporic acid as well as the plant hormones ABA and SLs (Alder et al. 2012; Medina et al. 2011; Schwartz et al. 1997, Fig. 1.6). The synthesis of these compounds is catalyzed by a ubiquitous family of non-heme iron enzymes called carotenoid cleavage dioxygenases (CCDs). CCDs utilize molecular oxygen to break C–C double bonds, splitting their substrates into two products that carry either an aldehyde or a ketone functional group. Members of this enzyme family show different substrate specificities with regard to the type and stereo-configuration (*cis/trans*) of the carotenoid substrate and the position of the double bond that is cleaved. In addition, some CCDs convert apocarotenoids instead of intact carotenoids. Carotenoid cleavage leads to a plentitude of products with different physicochemical properties, ranging from volatiles and scents with short-chain lengths, such as citral (C_{10}) in citrus fruits or β-ionone (C_{13}) in roses (Chhikara et al. 2018; Huang et al. 2009), to long-chain, lipophilic pigments, such as the fungal neurosporaxanthin and the citrus fruit pigment citraurin (C_{30}) (Estrada et al. 2008; Rodrigo et al. 2013).

There are five major plant CCD subfamilies: CCD1, CCD4, CCD7, CCD8, and the nine-*cis*-epoxy-carotenoid dioxygenases (NCEDs) (Auldridge et al. 2006; Walter and Strack 2011). NCEDs are responsible for ABA biosynthesis, catalyzing the stereospecific cleavage of 9-*cis*-violaxanthin and 9′-*cis*-neoxanthin at the C11–C12 (for carbon numbering of carotenoids, see Fig. 1.6) double bond to produce the ABA precursor xanthoxin (C_{15}) (Giuliano et al. 2003; Schwartz et al. 1997). CCD1 enzymes cleave different cyclic and acyclic all-*trans*-carotenoids as well as apocarotenoids at several positions, leading to a wide range of products. It has been assumed that the primary function of CCD1 is the scavenging of damaged carotenoids, allowing their replacement by intact ones (Ilg et al. 2010; Scherzinger and Al-Babili 2008). However, CCD1 has also been implicated in the synthesis of special pigments formed upon colonization of plant roots by symbiotic mycorrhizal fungi (Floss et al. 2008; Walter 2013). CCD4 enzymes catalyze the cleavage of the C7′–C8′ or the C9′–C10′ double bond in bicyclic all-*trans* carotenoids, leading to C_{10}- or C_{13}-volatiles and C_{30} or C_{27} apocarotenoids (Bruno et al. 2015, 2016; Ma et al. 2013; Rodrigo et al. 2013). The CCD4-mediated cleavage of the C9′–C10′ double bond in all-*trans*-bicyclic carotenoids determines the carotenoid content in different plant tissues, such as *Arabidopsis* seeds (Gonzalez-Jorge et al. 2013), *Chrysanthemum* flowers (Ohmiya et al. 2006), and potato tubers (Campbell et al. 2010; Bruno et al. 2015). In addition, the Arabidopsis CCD4 is supposed to mediate the formation of a hitherto unidentified signal from acyclic *cis*-configured desaturation intermediates, which regulates leaf and early chloroplast development (Avendano-Vazquez et al. 2014). CCD7 and CCD8 (Alder et al. 2012) are responsible for SL biosynthesis and will be discussed below.

1.2.3 Genetic Identification of Key SL Biosynthetic Genes

Matusova et al. (2005) demonstrated that carotenoids are the precursor of SLs. This conclusion was drawn from lack of SL activity in root exudates obtained from carotenoid deficient maize mutants and from maize, cowpea, and sorghum plants treated with carotenoid biosynthesis inhibitors. The presence of two moieties in SLs and their connection by an enol ether bridge led to the assumption that each of these moieties (ABC- and D-ring in canonical SLs; non-canonical SLs were unknown at that time) is synthesized separately, followed by a coupling reaction to form a parent SL molecule (5-deoxystrigol or 4-deoxyorobanchol). Because of lack of knowledge about the involved enzymes, Matusova et al. (2005) proposed that the ABC-ring, consisting of a C_{14} skeleton, arises from a C_{15}-apocarotenal (it is not possible to generate a C_{14}-apocarotenal from direct cleavage of a C=C double bond in carotenoids), which would subsequently be converted by a series of hypothetical reactions into a tricyclic lactone (ABC-ring) that is then linked in a final step to a D-ring of unknown origin. The C_{15}-apocarotenal was supposed to be xanthoxin, the precursor of ABA, or 9-*cis*-β-apo-11-carotenal that can be formed by cleaving 9-*cis*-β-carotene. The proposed biosynthesis route consisted of nine steps, implying a quite complicated biosynthetic pathway involving many enzymes.

At about the same time, plant developmental biologists from different research groups were trying to identify a putative plant growth regulator that is transported from roots to the shoot to limit the number of branches by inhibiting the outgrowth of axillar buds. The presence of this inhibitory signal was proposed upon the discovery of mutants with increased numbers of shoot branches/tillers and was supported by grafting studies showing that a wild-type rootstock can rescue the more branching/high-tillering phenotype of mutant scions. These mutants were called *more axillary growth* (*max*) in *Arabidopsis* (Booker et al. 2004, 2005; Sorefan et al. 2003; Stirnberg et al. 2007), *ramosus* (*rms*) in pea (Beveridge et al. 1996; Foo et al. 2005; Morris et al. 2001; Sorefan et al. 2003), *dwarf* (*d*)/*high-tillering dwarf* (*htd*) in rice (Arite et al. 2007, 2009; Ishikawa et al. 2005; Lin et al. 2009; Zhou et al. 2013), and *decreased apical dominance* (*dad*) in petunia (Drummond et al. 2009, 2011; Hamiaux et al. 2012; Simons et al. 2007; Snowden et al. 2005). Table 1.1 gives an overview of these mutants and the function of the corresponding genes. Mapping of the two mutants groups *max3*, *rms5*, *dad3*, and *d17/htd1* on the one hand, and *max4*, *rms1*, *dad1* and *d10* on the other, identified *CCD7* and *CCD8* as the respective corresponding genes, suggesting a carotenoid origin of the branching inhibitory signal. Identification of the loci corresponding to further mutants demonstrated that the synthesis of the inhibitory signal also required a cytochrome P450 of the clade 711 (MAX1 in Arabidopsis), a small iron-containing protein (DWARF27 in rice) and an oxoglutarate-dependent oxidoreductase (Lateral branching Oxidoreductase; LBO) in Arabidopsis) (Booker et al. 2005; Brewer et al. 2016; Lin et al. 2009). The breakthrough discovery that SLs are identical with the postulated inhibitory signal (Gomez-Roldan et al. 2008; Umehara et al. 2008) enabled deciphering the SL biosynthesis step by step, leading, in a relatively short time, to

Table 1.1 Genes involved in strigolactone (SL) biosynthesis and signaling (modified from Jia et al. 2018)

Role	Protein identity/function	Gene name			
		Arabidopsis	Pea	Petunia	Rice
SL biosynthesis	9-*cis*/all-*trans*-β-carotene isomerase	At*D27*	–	–	*D27*
	CCD7	*MAX3*	*RMS5*	*DAD3*	*HTD1/D17*
	CCD8	*MAX4*	*RMS1*	*DAD1*	*D10*
From CL to various SLs	Cytochrome P450, 711 (CYP711)	*MAX1*		*PhMAX1*	*CO (Os01g0700900), OS (Os01g0701400), Os02g0221900, Os06g0565100*
	Other downstream enzymes	*LBO*			
SL signaling — Perception	α/β hydrolase	At*D14*	*RMS3*	*DAD2*	*D14/D88/HTD2*
	F-Box protein	*MAX2*	*RMS4*	*PhMAX2A, PhMAX2B*	*D3*
Transcription regulation	Repressor of SL signaling	*SMXLs*			*D53*
	Corepressors	*TPR2*			*TPL/TPR2*
	Transcriptional factors	*BRC1, BRC2*			*IPA1*

CL carlactone, *CCD7* carotenoid cleavage dioxygenase 7, *CCD8* carotenoid cleavage dioxygenase 8, *MAX* more axillary growth, *RMS* ramosus, *DAD* decreased apical dominance, *HTD* high-tillering dwarf, *D* dwarf, *LBO* lateral branching oxidoreductase, *SMXL* SMAX1-LIKE, *TPL/TPR* topless/topless-related protein, *BRC* branching, *IPA* ideal plant architecture

an almost complete picture of the core of this pathway and showing that it needs much less enzymes than initially assumed.

1.2.4   From β-Carotene to Carlactone

Engineered *E. coli* strains that accumulate carotenoids are powerful tools to investigate the activity of carotenoid biosynthesis and modifying enzymes. These strains are usually generated through introducing carotenoid gene clusters from the bacteria *Pantoea agglomerans* and *Pantoea ananas* (formerly known as *Erwinia herbicola* and *Erwinia uredovora*, respectively), which confer *E. coli* cells with colors of accumulated all-*trans*-carotenoids. Transformation of these strains with CCDs leads to decolorization due to shortening of the chromophore (conjugated double-bond system) upon cleavage, given that the CCD can cleave the substrate produced by the particular strain. HPLC, GC-MS, and LC-MS analysis of carotenoids and apocarotenoids can subsequently identify substrate and product(s) of the introduced CCDs. Booker et al. (2004) used this in vivo system to characterize the activity of the Arabidopsis AtCCD7. The authors transformed the corresponding cDNA in phytoene, ζ-carotene, lycopene, δ-carotene (a monocyclic carotene that carries a ε-ionone ring), β-carotene, and zeaxanthin (for structures, see Fig. 1.5) accumulating *E. coli* cells. GC-MS analysis detected C_{13} volatiles as AtCCD7 products produced in β-, ζ-, δ-carotene and zeaxanthin strains. The formation of these volatiles implies that AtCCD7 cleaves the C9–C10 and/or C9′–C10′ double bond(s). This activity must also lead to a second product, a C_{27}-apocarotenal (plant carotenoids are C_{40} compounds), in case of a single cleavage reaction, and a C_{14}-diapocarotenal, in case of double cleavage. However, the authors did not provide data on the nature of the second product. Therefore, the question whether AtCCD7 catalyzes a single or double cleavage reaction remained unanswered. In the same year, Schwartz et al. (2004) published a study on the enzymatic activity of AtCCD7 and AtCCD8. The authors confirmed the AtCCD7 activity, in β-carotene- and lycopene-accumulating *E. coli* cells, which leads to C_{13}-volatiles. Moreover, they showed that the second product is a C_{27}-apocarotenal (β-apo-10′-carotenal in case of β-carotene accumulating *E. coli* cells), suggesting a single cleavage reaction. In addition, Schwartz et al. (2004) investigated the activity of AtCCD7 in vitro. They expressed the enzyme in normal *E. coli* cells, purified and incubated it with different carotenoid substrates. Assays were then analyzed by HPLC. The in vitro incubation with β-carotene isolated from spinach leaves led to the formation of a C_{27}-apocarotenal. In a second experiment, AtCCD7 and AtCCD8 were simultaneously introduced in β-carotene accumulating cells, followed by HPLC analysis of the apocarotenoid content. This analysis unraveled a C_{18}-apocarotenoid (β-apo-13-carotenone, also called D'orenone (Schlicht et al. 2008)) as a product of combined AtCCD7/AtCCD8 activity (Schwartz et al. 2004). This result indicated that AtCCD7 and AtCCD8 catalyze a sequential cleavage of all-*trans*-β-carotene into β-apo-13-carotenone and that AtCCD8 cleaves the AtCCD7 product β-apo-10′-carotenal at the C13–C14 double bond (for carbon numbering of carotenoids, see Fig. 1.6). Indeed, an in vitro study by Alder et al. (2008) showed that CCD8 enzymes from Arabidopsis and pea catalyze the conversion of all-*trans*-β-apo-10′-carotenal (C_{27}) into β-apo-13-carotenone (C_{18}). However, the authors did not identify the second product, a presumed C_9-dialdehyde that is expected to arise together with β-apo-13-carotenone (C_{18}). These results led to the assumption that β-apo-13-carotenone is the precursor of the shoot inhibitory signal that was later demonstrated to be SL. However, there were two reasons that raised doubts in this biosynthesis scheme. The first one is the many structural differences between β-apo-13-carotenone (C_{18}) and SLs, which would require an even more complicated pathway than the one proposed to start with a C_{15}-apocarotenal (see Sect. 1.2.3). The second reason was the failure of β-apo-13-carotenone to rescue the high-tillering phenotype of the rice *ccd8* mutant (*d10*) (Alder et al. 2012), which contradicts the presumed role of this compound as SL precursor. Hence, there was a need for revisiting CCD7 and CCD8 enzymatic activities.

1.2.4.1 Stereospecificity of CCD7

cis/trans isomerization is a major feature of carotenoids and apocarotenoids, which plays a crucial role in biological processes, such as light perception (also in vision!) and the formation of the plant hormone ABA that arises from a specific, *cis*-configured precursor (9-*cis*-violaxanthin or 9′-*cis*-neoxanthin) (Schwartz et al. 1997). However, this aspect had been overlooked in the course of investigating AtCCD7 and AtCCD8 activity. By virtue of conjugated double-bond systems, carotenoids and apocarotenoids are prone to photo- and thermo-induced *cis/trans* isomerization. This sensitivity impedes conclusions about the isomeric state of CCD substrates and products, if suitable analytical methods that can distinguish the *cis/trans* isomers are missing. For instance, carotenoid-accumulating *E. coli* strains contain traces of *cis*-configured carotenoids that arise by physical (photo- or thermo-) isomerization of the all-*trans* isomers formed by the introduced biosynthetic genes. In addition, β-carotene, isolated from plant material without further separation, is a mixture of different isomers, though all-*trans*-β-carotene is the major constituent. Taking these considerations into account, Alder et al. (2012) revisited the proposed AtCCD7 activity, by employing in vitro assays and using pure β-carotene stereoisomers. Analysis of incubations with CCD7 from *Arabidopsis*, rice, and pea demonstrated a clear substrate stereospecificity of these enzymes. They cleaved 9-*cis*-β-carotene, but not all-*trans*-, 13-*cis*-, or 15-*cis*-β-carotene. The cleavage reaction led to β-ionone and a β-apo-10′-carotenal isomer different from the all-*trans*-one, as demonstrated by HPLC analysis (Fig. 1.7). Hence, it was concluded that the β-apo-10′-carotenal produced by CCD7 is 9-*cis*-configured, which was supported by comparison to a tentative 9-*cis*-β-apo-10′-carotenal and later by NMR analysis (Alder et al. 2012; Bruno et al. 2014).

In vitro incubation with further 9-*cis*-configured carotenoids confirmed the strict stereo- and regiospecificity of the CCD7 enzymes. CCDs from *Arabidopsis*, pea, and

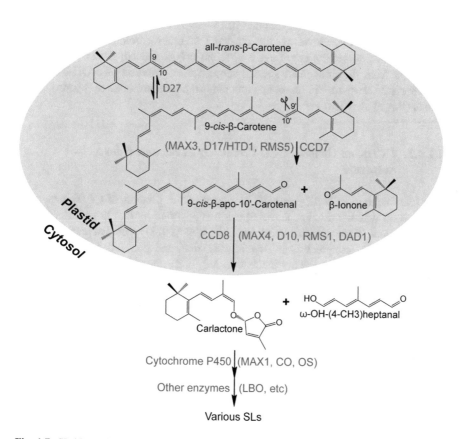

Fig. 1.7 SL biosynthesis. The 9-*cis*/all-*trans*-β-carotene isomerase D27 catalyzes the reversible conversion of all-*trans*-β-carotene (C_{40}) into 9-*cis*-β-carotene that is cleaved by the stereospecific carotenoid cleavage enzyme CCD7 at the C9′–C10′ double bond into 9-*cis*-β-apo-10′-carotenal (C_{27}) and β-ionone (C_{13}). In the third step, CCD8 converts 9-*cis*-β-apo-10′-carotenal into carlactone (C_{19}) and the C_8 compound ω-OH-(4-CH3)heptanal. The C-skeleton of carlactone corresponds to the shaded part of 9-*cis*-β-apo-10′-carotenal. D27, CCD7, and CCD8 are all localized in the plastids. Carlactone is the central metabolite of SL core pathway, which is catalyzed by cytochrome P450 enzymes, such as the *Arabidopsis* MAX1 and the rice carlactone oxidase (CO) and orbanchol synthase (OS), into carlactonoic acid and canonical SLs. Other enzymes, such as the lateral branching oxidoreductase are involved in later steps of SL biosynthesis, which contribute to SL diversity

the moss *Physcomitrella patens* converted 9-*cis*-zeaxanthin and 9-*cis*-lutein into 9-*cis*-3-OH-β-apo-10′-carotenal and 9-*cis*-3-OH-ε-apo-10′-carotenal, respectively (Bruno et al. 2014; Decker et al. 2017). An in vitro study of AtCCD7 activity on acyclic carotenoids led to similar results (Bruno et al. 2016). The enzyme converted 9-*cis*-ζ-carotene, 9′-*cis*-neurosporene, and 9-*cis*-lycopene, producing the corresponding 9-*cis*-configured products 9-*cis*-ζ-apo-10′-carotenal and 9-*cis*-apo-10′-lycopenal, respectively (Bruno et al. 2016). AtCCD7 and PsCCD7 showed higher affinity to 9-*cis*-β-carotene than to other substrates, such as 9-*cis*-zeaxanthin, which is consistent with the assumption that 9-*cis*-β-carotene is the precursor of SLs (Bruno et al. 2016). However, the cleavage of other carotenoids points to the possibility that some SLs may have a different origin. Indeed, a very recent study on CCD8 from different plant species showed the formation of 3-OH-carlactone from 9-*cis*-3-OH-β-apo-10′-carotenal and confirmed the presence of this product in plants (Baz et al. 2018).

1.2.4.2 CCD8, an Unusual CCD Producing Carlactone from a cis-Substrate

The identification of 9-*cis*-β-apo-10′-carotenal as the product of CCD7 raised the question on whether this apocarotenal can be converted by CCD8 and, if yes, into which product. To answer this question, Alder et al. (2012) incubated heterologously generated CCD8 enzymes from *Arabidopsis*, rice, and pea with CCD7-produced 9-*cis*-β-apo-10′-carotenal. This incubation led to an unexpected product that was called carlactone and which differed from known carotenoid/apocarotenoid cleavage products (Fig. 1.7). Carlactone showed surprisingly common structural features with SLs and is identical with them in the number of C atoms (C_{19}). It is a non-carbonyl, tri-oxygenated compound that already contains a lactone ring identical to the D-ring of SLs. Moreover, the two moieties of carlactone are connected by an enol ether bridge that is present in all SLs. Carlactone also shares with natural SLs the same stereo-configuration (*R*-configuration) at the C2′ atom, as shown later by Seto et al. (2014). Moreover, application of carlactone restored the wild-type tillering phenotype of the rice *ccd8* (*d10*) mutant and induced the germination of *Striga* seeds. These results suggested that carlactone is the intermediate of SL biosynthesis, which is formed by CCD8, although this compound was initially identified only as an in vitro product. Indeed, the presence of carlactone in rice and Arabidopsis was later demonstrated by a different group (Seto et al. 2014).

The formation of carlactone by CCD8 from 9-*cis*-β-apo-10′carotenal and the previously reported cleavage of all-*trans*-β-apo-10′carotenal into β-apo-13-carotenone suggest that this enzyme is an unusual CCD that catalyzes different types of reactions, depending on the stereochemistry of the substrate. When forming β-apo-13-carotenone, it is assumed that CCD8 utilizes a common dioxygenase mechanism, i.e., introduction of molecular oxygen into a C=C double bond, which leads to the formation of a dioxetane that breaks down into two carbonyl products (Fig. 1.8a), while the structure of carlactone implies different types of reactions and repeated oxygenations (Fig. 1.8b). To get insight into the reaction mechanism used by CCD8 to form carlactone, Bruno et al. (2017) applied a derivatization reagent to catch the second product expected to arise simultaneously with carlactone (C_{19}) from 9-*cis*-β-apo-10′-carotenal (C_{27}) and which had not been detected before. This approach identified ω-OH-(4-CH3)-hepta-2,4,6-trien-al as a further, unusual CCD product formed by CCD8, which carries an aldehyde and an alcohol functional group (Fig. 1.8B). The authors also used $^{18}O_2$ and ^{13}C labeling, which demonstrated that the three O atoms of carlactone derive from atmospheric

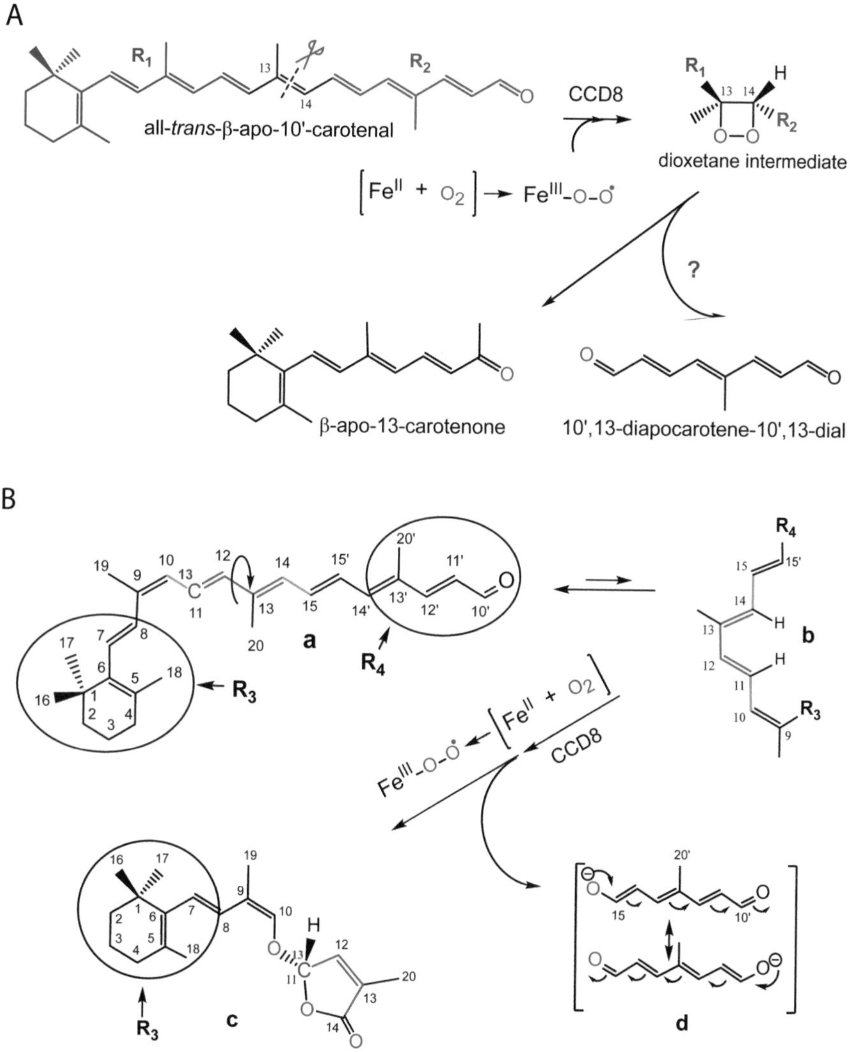

Fig. 1.8 A proposal for the mechanism of CCD8-catalyzed reactions (modified from Jia et al. 2018 and Bruno et al. 2017). (**a**) A proposal for the formation of β-apo-13-carotenone by CCD8 from all-*trans*-β-apo-10′-carotenal, the second substrate converted by CCD8 (though at much lower rate). R1 and R2 correspond to the structures present in the same color, respectively. The reactive Fe^{III}-O-O˙ species attacks the C14 of all-*trans*-substrate, forming an instable dioxetane intermediate at the C13–C14 double bond. The dioxetane breaks into two carbonyl products, β-apo-13-carotenone and a yet unconfirmed dialdehyde. (**b**) A simplified proposal for the mechanism of the formation of carlactone (c) from 9-*cis*-β-apo-10′-carotenal (a). The C numbering in b-d corresponds the C numbering in (a). R3 and R4 correspond to the β-ionone moiety and the C17–C21 chain of the substrate (a), respectively. Oxygen atoms are depicted in red. It is proposed that the formation of carlactone (c) is initiated (step I) by converting the *transoid* configuration of the bonds depicted in magenta into a cisoid one, by rotating at C12–C13 in (a). The attack of Fe^{III}-O-O˙ species, formed by the activation of atmospheric O_2 by the ferrous iron in CCD8 active

oxygen and that the C11 atom in carlactone (C2′ atom in SLs) corresponds to the C11 atom in 9-*cis*-β-apo-10′-carotenal. Based on these results, Bruno et al. (2017) developed a proposal for the mechanism of carlactone formation, which includes a series of reactions composed of isomerization, repeated oxygenation, and intramolecular rearrangements. Furthermore, it is proposed that the stereo-configuration of the C9–C10 double bond determines whether the enzymes build a dioxetane intermediate, leading to β-apo-13-carotenone, or a cyclic endoperoxide intermediate, to form carlactone (Bruno et al. 2017). However, further experimental and structural data are still needed to confirm this proposed mechanism.

1.2.4.3 Dwarf27, a 9-*cis*/all-*trans* β-Carotene Isomerase

The stereospecificity of CCD7 and CCD8 suggests that SL biosynthesis relies on 9-*cis*-β-carotene as precursor. 9-*Cis*-β-carotene is a natural compound that may be formed from all-*trans*-β-carotene, the isomer produced by the plant carotenoid pathway, by photo- or thermo-isomerization. However, Alder et al. (2012) assumed that hormone synthesis would need a more regulated precursor supply and, hence, that 9-*cis*-β-carotene is formed from the all-*trans* isomer by an enzymatic activity. Some years before, the iron-containing OsDWARF27 (OsD27) rice protein was identified as a plastidial enzyme required for SL biosynthesis, but with unknown enzymatic activity (Lin et al. 2009). The role of D27 in SL biosynthesis was also further confirmed by investigating the corresponding Arabidopsis mutant (Waters et al. 2012a). Hence, it was tempting to hypothesize that OsD27 might be the first described enzyme with an all-*trans*/9-*cis*-β-carotene isomerase activity. To test this hypothesis, Alder et al. (2012) expressed OsD27 in a β-carotene accumulating *E. coli* strain and analyzed the β-carotene pattern. This experiment showed that the introduction of OsD27 resulted in an increase in the 9-*cis*-/all-*trans*-β-carotene ratio, indicating an isomerase activity. In vitro assays performed with heterologously expressed and purified OsD27 confirmed this activity and showed that OsD27 catalyzes the isomerization of all-*trans*- into 9-*cis*-β-carotene and vice versa, resulting in an isomer equilibrium (Fig. 1.7). The mechanism of the D27-catalyzed β-carotene isomerization is still unknown. However, OsD27 activity is inhibited in the presence of silver acetate, indicating the involvement of an iron-sulfur cluster in the catalysis (Lin et al. 2009). OsD27 also isomerizes other carotenoids, such as the mono-hydroxylated β-carotene derivative β,β-cryptoxanthin (leading to 9-*cis*-β,β-cryptoxanthin) and α-carotene (leading to 9-*cis*-α-carotene and 9′-*cis*-α-carotene), which have at least one unmodified β-ionone ring (Bruno and Al-Babili 2016). While 9-*cis*-β, β-cryptoxanthin,

---

Fig. 1.8 (continued) center, at C14 in (b), followed by repeated oxygenation and intramolecular rearrangements, leads to the production of carlactone (c) and an extensively conjugated ω-OH-aldehyde group (d). C11 in carlactone (c) corresponds to C11 in 9-*cis*-β-apo-10′-carotenal (a), as confirmed by using accordingly ^{13}C-labeled 9-*cis*-β-apo-10′-carotenal (a)

9-*cis*-α-carotene, and 9-*cis*-β-carotene can be converted by CCD7 into the same SL biosynthesis intermediate 9-*cis*-β-apo-10′-carotenal, the question about the biological significance of 9′-*cis*-α-carotene produced by CCD7 remains open.

In both rice and Arabidopsis, *d27* mutants display less severe tillering/branching phenotype, compared to *ccd7* or *ccd8* mutants (Lin et al. 2009; Waters et al. 2012a). This difference might be explained by spontaneous isomerization of all-*trans*-β-carotene that can partially compensate lack of D27 activity. However, it is also possible that the D27 homologs (D27-LIKE 1 and D27-LIKE 2) present in Arabidopsis, rice, and other plant species also contribute to the formation of 9-*cis*-β-carotene. It should also be mentioned that 9-*cis*-β-carotene is not only a precursor of SLs but is also a structural component of the photosynthetic cytochrome *b6f* complex (Cramer et al. 2006).

1.2.5   From Carlactone to Canonical and Non-canonical SLs

Cytochrome P450 (CYP) enzymes are heme-containing monooxygenases that form a large enzyme superfamily in plants, animals, fungi, and bacteria. CYPs catalyze different types of reactions, such as C-hydroxylation, dealkylation, and epoxide formation (Isin and Guengerich 2007; Werck-Reichhart and Feyereisen 2000), and are involved in a wide range of biochemical pathways including lipids, alkaloids, terpenoids, and phenylpropanoids, as well as plant hormones (Chapple 1998; Werck-Reichhart and Feyereisen 2000).

The *Arabidopsis max1* mutant is disrupted in a gene encoding a CYP of the 711 clade (CYP711A1; MAX1) and shows a high-branching phenotype similar to that of *max2*, *max3*, and *max4* mutants (Booker et al. 2005; Stirnberg et al. 2002). Genetic analysis and reciprocal grafting experiments suggested that MAX1 catalyzes a reaction downstream of MAX3 (CCD7) and MAX4 (CCD8) to produce the shoot branching inhibitory signal (SLs, Booker et al. 2005). This order of reactions is supported by the plastid localization of D27, CCD7, and CCD8 that act on carotenoids/apocarotenoids precursor(s) present in plastids and by the cytosolic localization of MAX1. The similarity of carlactone to SLs suggested that only a few steps would be required to convert carlactone into real SLs (Alder et al. 2012). In the absence of further candidate enzymes, it was assumed that carlactone may be a direct substrate of MAX1. Supporting this assumption, it was reported that the content of carlactone in the *Arabidopsis max1* mutant is approximately 700-fold higher than in wild-type plants (Seto et al. 2014). In addition, feeding experiments of *max4* and *max1max4* mutants with ^{13}C-labeled carlactone confirmed the role of MAX1 in carlactone conversion. The direct evidence for the role of MAX1 in this process was provided by in vitro studies using yeast as expression system. Incubation of yeast microsomes expressing MAX1 with carlactone showed the consecutive oxidations of the C19 atom, leading to a molecule designated by Abe et al. (2014) as carlactonoic acid (CLA) (Fig. 1.9). The authors confirmed the in vitro results by feeding *max4* and *max1max4* mutants with ^{13}C-labeled carlactone, which showed the formation of ^{13}C-labeled CLA in *max4* but not in *max1max4* double mutant. CLA is an intermediate in the

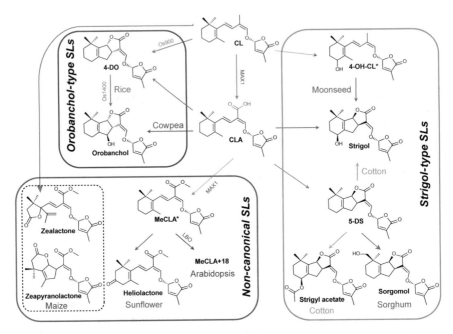

Fig. 1.9 Pathways from carlactone to SLs in different plant species. Names of the species are displayed in different colors, and names of enzymes from the same species are indicated in the corresponding color. *Intermediacy of 4-OH–CL in moonseed and MeCLA in sunflower has not been confirmed. Abbreviations: CL, carlactone; CLA, carlactonoic acid; MeCLA, methyl carlactonoic acid; 4-DO, 4-deoxyorobanchol; 4-OH–CL, 4-hydroxy carlactone; 5-DS, 5-deoxystrigol

Arabidopsis SL biosynthesis pathway and is methylated by an unidentified methyltransferase into methyl carlactonoic acid (MeCLA) (Fig. 1.9, Abe et al. 2014).

Recently, transcriptomic studies led to the identification of a further Arabidopsis SL biosynthesis enzyme. The enzyme *LATERAL BRANCHING OXIDOREDUC-TASE (LBO)*, a member of the 2-oxoglutarate and Fe^{II}-dependent dioxygenase family, is co-expressed with MAX3 (Brewer et al. 2016), and an *lbo* loss-of-function mutant showed a more branching phenotype, which is, however, less pronounced than in *max3* and *max4* mutants. Further grafting and genetic analysis suggested that LBO acts downstream of MAX1 in the SL biosynthetic pathway (Brewer et al. 2016). Moreover, investigation of the activity of heterologously produced LBO in vitro showed the conversion of MeCLA into an unidentified product with a molecular weight equal to MeCLA+16 Da, suggesting the introduction of an oxygen atom in this substrate (Fig. 1.9, Brewer et al. 2016).

The rice genome encodes five *MAX1* homologs including a truncated, nonfunctional one (Challis et al. 2013). Zhang et al. (2014) investigated the activity of the four functional rice MAX1 homologs, aiming at the identification of carlactone-metabolizing enzymes. For this purpose, the authors expressed the four MAX1 enzymes in yeast and isolated the corresponding microsomes for in vitro

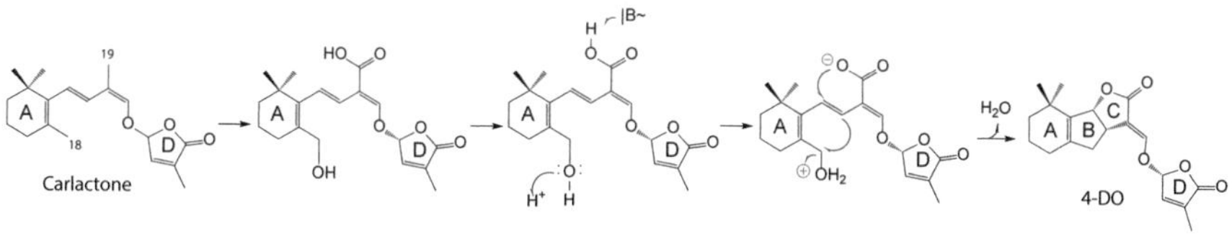

Fig. 1.10 Proposed mechanism for the formation of 4-deoxyorobanchol (4-DO) from carlactone oxidase (CO) (adapted from (Zhang et al. 2014)). The enzyme catalyzes three oxygenation reactions, with carboxylation of the C19, and hydroxylation of the C18 atom. Abstraction of a proton from the carboxyl group by an active center base (~B|) results in the formation of C-ring, and synchronously the addition of a proton to the hydroxyl group at C18 triggers the stereospecific B-ring closure and the release of water to form the canonical SL, 4-DO

studies utilizing carlactone as substrate (Zhang et al. 2014). In addition, they used the *Nicotiana benthamiana* transient expression system, by simultaneous introduction of genes encoding the carlactone biosynthetic genes alone, as a control, and together with each of the functional rice *MAX1s*. Leaves of transformed *N. benthamiana* plants were then analyzed by LC-MS to determine carlactone conversion and identify any product (s) formed. The two approaches led to the identification of the rice MAX1 homolog Os900 (carlactone oxygenase, CO) as the enzyme that converts carlactone into 4-deoxyorobanchol (4DO; or *ent-2′-epi*-5-deoxystrigol), the parent molecule of the orobanchol-type, canonical SLs (Zhang et al. 2014). The conversion of carlactone into 4DO possibly involves a series of reactions resulting in oxidation, ring closure, and B/C lactone moiety formation, all catalyzed by Os900 (Figs. 1.9 and 1.10). Intriguingly, simultaneous transient expression of *Os900*, *Os1400*, *Os5100*, and *Os1900* with the carlactone biosynthetic genes caused a significant decline in 4DO and an increase in orobanchol, compared with co-expression of *Os900* (Zhang et al. 2014). This suggested that one of the other MAX1s (*Os1400*, *Os5100*, or *Os1900*) oxidizes 4DO to form orobanchol. Further analysis in *N. benthamiana* and in vitro assays using yeast microsomes showed that Os1400 is an orobanchol synthase that hydroxylates 4DO to form orobanchol (Fig. 1.9, Zhang et al. 2014).

Tomato, like Arabidopsis, contains only one CYP711A1 (SlMAX1, Zhang et al. 2018). A *Slmax1* mutant showed several SL deficiency phenotypes in plant architecture and development and displayed significantly reduced production of SLs. Transient expression of *SlMAX1* together with *SlD27*, *SlCCD7*, and *SlCCD8* in *N. benthamiana* leaves demonstrated that SlMAX1 catalyzes the conversion of carlactone to CLA, just as *Arabidopsis* MAX1. Intriguingly, in contrast to the situation in rice, orobanchol (one of the major SLs in tomato) and its direct precursor 4DO were not directly produced from carlactone by SlMAX1. Plant feeding assays indicated that CLA, but not 4DO, is a precursor of orobanchol, which in turn is the precursor of the other known SLs in tomato (Zhang et al. 2018).

The investigation of the activity of AtMAX1 and homologs from rice and tomato unraveled differences in their substrates and formed products. However, feeding of fluridone-treated plants from different species, including sorghum, cotton, moonseed, cowpea, and sunflower, with ^{13}C-labeled carlactone resulted in the formation of ^{13}C-CLA, indicating that the synthesis of CLA is a common step in SL biosynthesis (Fig. 1.9, Iseki et al. 2018). A recent, comparative study on AtMAX1 and homologs from rice, maize, tomato, poplar, and the lycophyte *Selaginella moellendorffii*, which used different heterologous systems to determine the substrate specificities, revealed the presence of three groups. The first one, called A1-type MAX1s, includes AtMAX1 and homologs from *Arabidopsis*, tomato, and poplar, which convert carlactone into CLA. A2-type MAX1s, represented by the rice Os900 and the *Selaginella* SmMAX1a/b, produce 4DO from carlactone. The A3 type of MAX1s, which includes the maize the maize ZmMAX1b and the rice Os1400, forms both CLA and 4-deoxyorobanchol from carlactone in vitro (Yoneyama et al. 2018). However, it can be speculated that A3-type MAX1s may have different catalytic activities in vivo. For instance, there is some evidence for the formation of orobanchol from 4-deoxyorobanchol by Os1400 in planta.

1.2.6 *The Biological Significance of Diversification in SLs*

So far, around 25 different SLs have been characterized in different plant species, and even a single plant species can usually produce many different SLs (Abe et al. 2014; Awad et al. 2006; Charnikhova et al. 2017; Kohlen et al. 2013; Xie 2016; Yoneyama et al. 2015). This is likely a consequence of the different biological functions exerted by SLs and indicates that structurally diversified SLs might be active in some biological functions but not in others (Wang and Bouwmeester 2018). Moreover, there is likely a selection pressure for plants producing SLs with minor alterations in the structure, which could maintain the interactions with AM fungi but avoid the infestations of parasitic weeds. Indeed, different parasitic species show divergent responses to the root exudates from different host species. For instance, sorgomol, a SL found in root exudates of sorghum, is much more active in inducing seed germination in *Striga hermonthica*, compared to *Orobanche minor* (Xie et al. 2008). *Ent-2'-epi-orobanchol* and its acetylated derivative are two SLs found in both cowpea and red clover. It has been shown that acetylated *ent-2'-epi-orobanchol* has higher germination-inducing activity toward *Orobanche minor* and *Striga hermonthica* than its non-acetylated form (Ueno et al. 2011), suggesting that a small change in SL structure can influence its biological activity in inducing seed germination in root parasitic plant. Worth to mention that Samodelov et al. developed a rapid and efficient biosensor with high sensitivity and specificity to different SLs, which may facilitate understanding the biological significance of SL diversity (Samodelov et al. 2016).

Recently, *LGS1* (*LOW GERMINATION STIMULANT 1*), which encodes an enzyme annotated as a sulfotransferase, was identified as a gene that determines the stereochemistry of canonical SLs released by sorghum roots (Gobena et al. 2017). Functional loss of *LGS1* altered the pattern of released SLs in sorghum, replacing the major SL 5-deoxystrigol by orobanchol. This alteration led to increased resistance to *Striga*, as 5-deoxystrigol is a more potent inducer of *Striga* seed germination than orobanchol (Gobena et al. 2017). Most intriguingly, this alternation in SL composition did not affect the other essential SL biological

functions, such as the capability to promote AM fungi colonization and the inhibition of tillering (Gobena et al. 2017). Although the mechanism underlying the effects
of LGS1 is still elusive, this discovery demonstrates the possibility of changing the
pattern of SLs and opens up new possibilities for generating resistance to *Striga*.

1.3 Regulatory Control of SL Production

1.3.1 Regulation by Nutrient Availability

During evolution, plants have acquired the ability to sense the availability of soil
nutrients and to respond accordingly. Nitrogen and phosphorus are two major
nutrients required for plant growth and development; however, both are quite
limiting nutrients in soil due to their low availability or low mobility. Therefore,
plants modulate their root architecture and recruit symbiotic partners, such as AM
fungi to increase the soil volume available for nitrogen and especially for phosphate
uptake (Gutjahr 2014). SLs have a critical role in both processes. Accordingly, SL
biosynthesis and production are promoted when nitrogen and particularly when
phosphate supply are insufficient, as shown for rice (Jamil et al. 2011; Umehara
et al. 2010), *Arabidopsis* (Kohlen et al. 2011), sorghum (Jamil et al. 2013), maize
(Jamil et al. 2012), and red clover (Yoneyama et al. 2007). Similarly, root exudates
of tomato plants grown under low-phosphate conditions showed increased activity in
inducing seed germination in *Orobanche ramosa*, compared to control plants
(Lopez-Raez et al. 2010). LC-MS/MS analysis confirmed the increased production
of several SLs, including orobanchol, solanacol, and didehydro-orobanchol isomers
under phosphate starvation. Similarly, the release of heliolactone by sunflower roots
is induced upon phosphate starvation (Ueno et al. 2014). Recently, the effect of
phosphate availability on germination stimulating activity of *Physcomitrella patens*
exudate was examined also using *O. ramosa* seeds (Decker et al. 2017). It was
demonstrated that SLs are released by *Physcomitrella* as confirmed by seed germination bioassay and that this release is increased by phosphate deficiency indicating
the evolutionarily conserved role of SLs as well as their regulation by phosphate
availability in plants (Decker et al. 2017).

The upregulation of SL content and release upon nutrient deficiency and during the
AM symbiosis is mainly a result of increased transcript levels of SL biosynthesis
genes, as shown for rice, *Medicago truncatula*, and tomato (Bonneau et al. 2013; Sun
et al. 2014; Wen et al. 2016; Stauder et al. 2018). Accordingly, sufficient phosphate
availability reduces transcript levels of SL biosynthesis and transporter genes, as
shown for petunia *DAD1/CCD8* and the SL transporter gene *PDR1* (Breuillin et al.
2010; Kretzschmar et al. 2012). Recently, it was shown that even the supply of
carotenoid precursors (β-carotene) for SL production is upregulated by phosphate
starvation and during root colonization by AM fungi. This supply is mediated by a
distinct symbiosis-inducible isoform of phytoene synthase, and precursor supply can
be limiting for SL production in dicot roots (Stauder et al. 2018).

1.3.2   SL Homeostasis and Regulation by Other Hormones

Like other plant hormones, SL biosynthesis is regulated within complex networks built by other hormones. In addition, SL biosynthesis is governed by a negative feedback mechanism maintaining SL homeostasis. For instance, application of the SL analog GR24 to Arabidopsis plants led to a decrease of *CCD7* and *CCD8* transcript levels (Mashiguchi et al. 2009). Consistently, SL biosynthesis and perception mutants in *Arabidopsis*, pea, rice, and petunia all showed elevated transcript levels of SL biosynthetic enzymes. This increase was observed with *CCD8* transcript in pea, rice, and petunia and with *MAX1*, *CCD7*, *CCD8*, and *LBO* transcripts in *Arabidopsis* (Arite et al. 2007; Brewer et al. 2016; Hayward et al. 2009; Simons et al. 2007; Snowden et al. 2005). In rice, the negative feedback signal is transduced via the SL signaling repressor D53, as shown by a gain of function *d53* mutant that displayed increased transcript levels of SL biosynthesis genes and elevated SL contents (Jiang et al. 2013; Yao et al. 2016).

The plant hormone auxin is a major regulator of plant growth and development. The feedback inhibition of SL biosynthesis is dependent on auxin levels and signaling. On the other hand, auxin is a major regulator of SL biosynthesis at the transcriptional level, as shown for pea, rice, Arabidopsis, Chrysanthemum, and tomato. In pea, removing the auxin source by decapitation or reducing polar auxin transport (PAT) by 1-N-naphthylphthalamic acid (NPA) treatment, an auxin transport inhibitor, significantly decreased the transcript levels of *RMS5* and, particularly, *RMS1* in the upper part of stems, which could be restored by auxin application (Foo et al. 2005; Johnson et al. 2006). Similar results were reported for transcript levels of *DgD27* in Chrysanthemum and *MAX3* in Arabidopsis (Hayward et al. 2009). Auxin also positively regulates the expression of *MAX4* and *D10* in Arabidopsis and rice, respectively (Arite et al. 2007; Wen et al. 2016). In *Arabidopsis*, the induction of *MAX3* and *MAX4* transcripts by auxin is dependent on *AUXIN RESISTANT 1* (*AXR1*), a subunit of the RUB1-activating enzyme that regulates auxin receptor complex (Hayward et al. 2009). In tomato, silencing of the gene encoding the auxin signaling component AUX/IAA protein Sl-IAA27 resulted in downregulation of SL biosynthetic genes, *D27* and *MAX1* (Guillotin et al. 2017).

ABA is an important plant hormone regulating various processes in plants and is mainly known for its role in the response to abiotic stresses (Sah et al. 2016; Vishwakarma et al. 2017). Several studies have shown that SLs are involved in stress response. In *Lotus*, osmotic stress decreased the SL content in tissues and root exudates by reducing transcript levels of SL biosynthetic and transporter genes (Liu et al. 2015). In tomato, it was shown that inhibition of NCED, a key enzyme in ABA biosynthesis, by applying AbaminSG decreased SLs levels, indicating that ABA maybe a positive regulator of SL biosynthesis (Lopez-Raez et al. 2010). Accordingly, ABA-deficient mutants, such as *notabilis*, showed reduced transcript levels of the SL biosynthetic genes *LeCCD7* and *LeCCD8* and released lower SL amounts (Lopez-Raez et al. 2010). A positive role of ABA in regulating SL biosynthesis in maize is indicated by decreased SL content in root exudates of the ABA-deficient mutant *viviparous14* (Lopez-Raez et al. 2010). In addition, recent studies indicated interference between ABA and SL biosynthesis in both rice and Arabidopsis, which is caused by an involvement of the SL biosynthesis enzyme D27 in determining ABA level (Abuauf et al. 2018; Haider et al. 2018). Analysis of *Osd27* mutant unraveled lower ABA content shoot and higher susceptibility to drought, compared to wild type (Haider et al. 2018).

Gibberellins (GAs) are plant hormones that regulate a series of developmental processes, such as germination, stem elongation, dormancy, and flowering. A recent investigation of the effect of GAs on SL biosynthesis in rice suggested a role as a negative regulator of SL biosynthesis. Treatment with GA1, GA3, or GA4 resulted in decreased 4-deoxyorobanchol levels in root tissues and exudates. The decrease in SL biosynthesis is caused by downregulation of *D27*, *D10*, *D17*, *Os900*, and *Os1400* transcript levels (Ito et al. 2017).

1.4 SL Transport and Exudation

Grafting is a commonly used technique for investigating metabolite and hormone fluxes between shoots and roots. SL-deficient mutants display an obvious "more branching/high-tillering" phenotype. However, the "bushy" phenotype could be fully rescued when mutant scions were grafted onto WT rootstocks, indicating that SLs are transported from roots to shoots (Kohlen et al. 2010). On the other hand, grafting wild-type scions onto rootstocks of SL-deficient mutants did not lead to increased branching, which suggests that SLs can be synthesized in aerial tissues as well (Kohlen et al. 2010). Grafting studies also indicated that SL biosynthesis intermediates can be transported from roots to shoots. Grafting of *max3* or *max4* scions onto *max1* rootstocks restored wild-type branching to *max3* and *max4* shoots, which suggests that carlactone or its derivative(s) accumulated in *max1* rootstocks is transported into *max3* or *max4* carlactone-deficient shoots where it can be further metabolized into SLs that inhibit the shoot branching (Booker et al. 2005).

The question on how SLs are transported from roots to shoots is still largely elusive. It has been assumed that SLs are transported through the xylem (Kohlen et al. 2010). However, upon feeding rice roots with labeled orobanchol and 4-deoxyorobanchol, both compounds were detected in the shoot, but not in the xylem sap (Xie et al. 2015). This experiment provided further evidence in favor of a different transport mechanism. Time kinetic studies showed that the transport of labeled SLs was much slower than expected for xylem-mediated movement (Xie et al. 2015). Thus, further studies are still required to understand the transport of SLs between organs.

In addition to root/shoot transport, SLs are released into the rhizosphere to facilitate the symbiosis with AM fungi, which is supported by the presence of specific transporters. In petunia, *P. hybrida PLEIOTROPIC DRUG RESISTANCE 1* (*PhPDR1*), a member of the ATP-binding cassette (ABC) transporter superfamily, was shown to mediate the release of SLs from the roots into the soil (Kretzschmar et al. 2012). Localization studies indicated that *PhPDR1* is predominantly present in the plasma membrane of individual subepidermal cells in lateral roots which largely correspond to hypodermal passage cells that act as entry points for AM fungi hyphae. This may suggest that the PhPDR1 activity leads to a local SL maximum guiding the hyphae of AM fungi hyphae to the hypodermal passage cells. *PaPDR1*, a *PhPDR1* ortholog from *P. axillaris*, exhibits a cell-type-specific asymmetric localization in different root tissues (Sasse et al. 2015). This transporter is localized at the apical membrane of root hypodermal cells in root tips, which may explain the shootward transport of SLs. Above the root tips, in the hypodermal passage cells, PaPDR1 is localized in the outer-lateral membrane, consistent with its postulated function to transport SLs from the roots into the rhizosphere.

1.5 SL Perception and Signaling

1.5.1 Perception of SLs

Perception and signal transduction of various plant hormones including SLs share a number of common features, of which the regulation via the ubiquitin 26S proteasome system (UPS) of the receptor protein complex including transcriptional repressors is the most-intensely studied part (Hershko and Ciechanover 1998). This main proteolytic pathway in eukaryotic cells is involved in ABA, auxin, brassinosteroid, cytokinin, ethylene, GAs, and jasmonic acid and also in SL perception (Dharmasiri et al. 2005; Jiang et al. 2013; Katsir et al. 2008; Kepinski and Leyser 2005; Mockaitis and Estelle 2008; Santner and Estelle 2009; Yao et al. 2016). UPS sequentially utilizes three types of ligases, E1, E2, and E3, to attach multiple units of ubiquitin to protein substrates, marking them for degradation by the 26S proteasome. There are five types of E3 ligases, including the SCF (SKP1, CULLIN, and F-BOX) type that plays an essential role in the signal transduction of plant hormones. SCF E3 ligase is a complex consisting of a conserved SKP1, a CULLIN protein, and a specific F-BOX protein that determines the substrates recognized by the E3 ligase (Shu and Yang 2017). In Arabidopsis, the F-box protein MAX2 (rice ortholog is DWARF3, D3) is a signal transduction component for both SLs and karrikins. Accordingly, *max2* mutants show all phenotypes related to SL deficiency but also reduced photomorphogenesis and decreased seed germination, two karrikin-regulated processes (Shen et al. 2012; Waters et al. 2014).

However, a special feature of SL perception is the involvement of DWARF14 (D14), a α/β-fold hydrolase that binds and catalyzes the hydrolysis of the hormone signal, in this case of SLs (Fig. 1.11) (Hamiaux et al. 2012; Yao et al. 2016). The capability to hydrolyze the ligand is an intriguing feature of D14, which distinguishes it from other hormone receptors, such as the gibberellin receptor GID1 (Ueguchi-Tanaka and Matsuoka 2010), and which has been demonstrated to be required for transducing the SL signal (Yao et al. 2016). SL hydrolysis is mediated by a conserved Ser/His/Asp catalytic triad and results in the formation of 5-hydroxy-3-methylbutenolide (D-ring) and a tricyclic lactone. However, when applied to

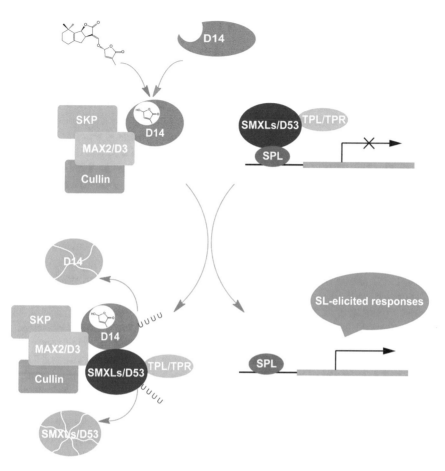

Fig. 1.11 SL perception and signaling pathway. In the absence of SL, the transcriptional repressors SMXLs/D53 together with the corepressor TPL/TPR interact with SPL transcription factors, inhibiting their transcriptional activity and repressing the expression of target genes. Binding of SL to the receptor D14 leads to the hydrolysis of the ligand, which releases the SL D-ring that covalently binds to the receptor. This process is accompanied by a simultaneous open-to-closed state conformational change that facilitates the interaction with the F-box protein D3 or MAX2. CLIM-D14-D3 further interacts with SMXLs/D53 to form a ternary complex, which will result in the polyubiquitination of D14 and SMXLs/D53 and, finally, in their degradation by the 26S proteasome. The degradation of D14 could desensitize SL signaling, and the degradation of SMXLs/D53 will relieve the repression of SPL by SMXLs/D53-TPL/TPR, which activates SL-elicited responses

plants, none of the two products showed any SL activity, suggesting that the purpose of SL hydrolysis is not to release an active final SL product. Hence, it was assumed that a conformational change in D14, which accompanies hydrolysis, is critical for SL signal transduction. Indeed, the SL analog GR24 promotes the physical interaction between D14 and MAX2 and triggers the consequent degradation of D14 by

MAX2, which require the active Ser/His/Asp catalytic triad in D14 protein (Hamiaux et al. 2012).

The crystal structure of the SL-induced rice AtD14-D3-ASK1 complex explained how the conformational change of AtD14 mediates SL signal transduction (Yao et al. 2016). Following binding and hydrolysis, the D-ring of SLs forms a covalently linked intermediate molecule (CLIM) with the receptor D14, while D14 undergoes an open-to-closed state conformational change to facilitate the interaction with D3 (Fig. 1.11) (Yao et al. 2016). Furthermore, it was demonstrated that the Ser in the Ser/His/Asp catalytic triad is responsible for docking of SLs into D14 (Zhao et al. 2013). The four helices near the active site pocket of D14 are also important for its activity, as corresponding mutations resulted in SL insensitivity due to a reduction in the aperture size of the active site pocket (Kagiyama et al. 2013; Nakamura et al. 2013). The formation of CLIM by D14 was also shown for pea D14 (de Saint Germain et al. 2016) and the *Striga* SL receptor ShHTL7 (Yao et al. 2017).

Karrikin-Insensitive2 (*KAI2*) and *DWARF14-LIKE2* (*DLK2*) are two close homologs of *AtD14* in *Arabidopsis* (Waters et al. 2012b). Despite its high similarity to D14, KAI2 shows different ligand specificity, binding karrikin and mediating the plant response to this compound that regulates photomorphogenesis and leaf development (Nelson et al. 2010; Soundappan et al. 2015; Waters et al. 2013). KAI2 has the conserved Ser/His/Asp triad that is required for karrikin response and possesses a smaller ligand-binding pocket than D14, indicating that this receptor recognizes ligands smaller than SLs (Zhao et al. 2013). However, the crystal structure of KAI2-KAR1 (KAR1 is a karrikin) complex indicates that KAI2-KAR1 creates a contiguous interface that allows the binding of signaling partners in a ligand-dependent manner and likely does not transduce the ligand signal by hydrolysis and CLIM formation, in contrast to D14 (Guo et al. 2013). There is not much information about the biological functions of DLK2, because Arabidopsis *dlk2* mutants do not show any obvious phenotype (Waters et al. 2012b). Root parasitic plants contain D14 paralogs termed *Striga hermonthica* HYPO-SENSITIVE TO LIGHT/KARRIKIN INSENSITIVE2 proteins (ShHTLs/ShKAI2s) that perceive SLs to induce seed germination (Conn et al. 2015; Tsuchiya et al. 2015). By experimental evidences, it was demonstrated that ShHTL7, similar to D14, is a non-canonical receptor that hydrolyzes SLs and generates a CLIM, which enhances the interaction between ShHTL7 and ShMAX2 to trigger *Striga* seed germination (Yao et al. 2017).

1.5.2 Transduction of the SL Signal

As described above, SCF E3 ligase-regulated UPS plays a central role in the signal transduction of plant hormones. The transduction of the SL signal uses the SCF^{MAX2} complex that contains the F-box protein MAX2 and the SCF subunits ASK1 and AtCUL1. A study of a rice high-tillering gain-of-function mutant led to the identification of SMXLs/D53 proteins as the $SCF^{MAX2/D3}$ substrates in the SL signaling pathway (Fig. 1.11) (Jiang et al. 2013; Yao et al. 2016). In rice, D14 directly interacts with D53 and D3 in a SL-dependent manner, leading to polyubiquitination and degradation of D53. A small deletion in the C-terminal of D53 stabilizes this protein, preventing its degradation, which results in phenotype similar to those of SL-deficient mutants (Jiang et al. 2013; Yao et al. 2016). This mutation and the resulting phenotype suggest that D53 is a negative regulator of SL signaling. In *Arabidopsis*, the D53 orthologues, SMXL6, SMXL7, and SMXL8, also interact with AtD14 and MAX2 and are negatively regulating shoot branching (Wang et al. 2015). There are eight members in the SMXLs/D53 family in rice and Arabidopsis, respectively (Stanga et al. 2013). It is assumed that SCF$^{MAX2/D3}$ may have different substrate preferences and hence regulate different SL responses. In *Arabidopsis*, it was shown that SMAX1 and SMXL2, two members of the SMXLs/D53 protein family, are involved in SL-independent processes, i.e., karrikin-dependent photomorphogenesis and seed dormancy (Soundappan et al. 2015; Stanga et al. 2013). Moreover, SMXL3, SMXL4, and SMXL5 were demonstrated to regulate phloem development, independent of both SLs and karrikins (Wallner et al. 2017). Sequence analysis showed that all SMXLs/D53 proteins contain a conserved D2 ATPase domain (Zhou et al. 2013). This D2 domain includes a conserved ethylene-responsive element binding factor-associated amphiphilic repression motif of Phe-Asp-Leu-Asn-Leu, which has been assumed to interact with TOPLESS/TOP-LESS RELATED PROTEIN2 (TPL/TPR2) (Fig. 1.11) (Ke et al. 2015; Pauwels et al. 2010; Soundappan et al. 2015; Zhou et al. 2013). Members of the TPL/TPR2 family usually act as transcription corepressors that inhibit the expression of genes related to plant hormones by repressing the transcriptional activity of interacting transcriptional factors (Ke et al. 2015). Indeed, it was also shown that D53, SMXL6, and SMXL8, all directly interact with TPL or TPR2 in vivo (Wang et al. 2015), suggesting that SMXLs/D53 proteins likely act as transcriptional repressors in SL signal pathway.

1.5.3 Downstream Transcription Factors in SL Signaling

Although the D3–D14-SMXLs/D53 regulatory module has been well characterized, the direct downstream transcriptional factors that regulate SL responsive gene expression are still largely unknown. One class of transcription factors (TFs), termed the *TEOSINTE BRANCHED1/CYCLOIDEA/PROLIFERATING CELL FACTOR1* (*TCP*) family, has been shown to act downstream of *SMXLs/D53* in SL signaling, such as *FC1/OsTB1* in rice (Takeda et al. 2003), *BRC1* and *BRC2* in Arabidopsis (Aguilar-Martinez et al. 2007), *TB1* in maize (Doebley et al. 1997; Guan et al. 2012), and *PsBRC1* in pea (Braun et al. 2012). These genes are mainly expressed in axillary buds, and corresponding loss-of-function mutants showed an increased branching/tillering phenotype that could not be recovered by exogenous SL application. In Arabidopsis, the expression level of *BRC1* was shown to be decreased in *max3* and *max2* but enhanced in *smxl6/7/8* (Soundappan et al. 2015; Wang et al. 2015). Similarly, expression of *HB53*, one of the known target genes activated by *BRC1*, was also increased in *smxl6/7/8* triple mutants (Wang et al. 2015). Moreover, the expression of *BRC1* and *PsBRC1* was induced by SL treatment in Arabidopsis and pea, respectively (Braun et al. 2012). However, in monocots, such as rice and maize, SL signaling and the expression of TB1 seem to be uncoupled; the expression of *FC1/OsTB1* and TB1 is not induced by SLs (Guan et al. 2012; Minakuchi et al. 2010). In addition, even though *BRC1* works downstream of *SMXLs/D53* in Arabidopsis SL signaling, there is no direct interaction between BRC1 and members of the SMXLs/D53 family. Hence, the connection between *SMXLs/D53* and *BRC1* is still missing.

Ideal Plant Architecture 1 (*IPA1*) (Jiao et al. 2010; Miura et al. 2010), a *SQUAMOSA PROMOTER BINDING PROTEIN-LIKE (SPL)* transcription factor (Wang and Wang 2015), is a key regulator of the plant architecture in rice that triggers the expression of *D53* (Song et al. 2017). Interestingly, IPA1 also interacts physically with D53, as shown by in vivo and in vitro studies. This interaction suppresses the transcriptional activity of IPA1, generating a negative feedback loop that controls *D53* expression, which is regulated by SLs (Song et al. 2017). IPA1 is the first example that shows how the SMXLs/D53 transcriptional repressors directly regulate downstream transcriptional factors (Fig. 1.11).

1.6   Conclusion

Strigolactones (SLs) are carotenoid-derived plant hormones and rhizosphere signaling molecules mediating the communication with symbiotic mycorrhizal fungi, which have been coopted by root parasitic weeds as seed germination signal.

The core pathway of SL biosynthesis is strictly stereospecific and leads from all-*trans*-β-carotene to the central intermediate carlactone, catalyzed by the sequential action of three enzymes, the all-*trans*/9-*cis*-β-carotene isomerase (D27), the 9-*cis*-specific CCD (CCD7), and the unusual CCD, CCD8, which catalyzes a combination of reactions on the 9-*cis*-product of CCD7.

CYP enzymes of the clade 711 (the *Arabidopsis* MAX1 and homologs) catalyze the conversion of carlactone to various canonical and non-canonical SLs, including carlactonoic acid, a supposed intermediate in the synthesis of both SL classes.

Plants produce different SLs, as observed in exudates of single plants. In addition, different plant species produce diverse SL compositions.

A set of largely unknown enzymes, including the recently identified oxoglutarate-dependent oxidoreductase, LBO, is responsible for the structural diversity of SLs. The elucidation of the biosynthesis routes behind this diversity will require combining multiple omic approaches, i.e., genomics, transcriptomics, and targeted metabolomics.

SL biosynthesis and release are regulated by the availability of nutrients, mainly phosphate, and governed by other plant hormones.

SL perception and signal transduction proceed via the ubiquitin 26S proteasome-based pathway that is used by other plant hormones. An SCF complex composed of

Skp, Cullin, and the F-box protein (MAX2 or D3) in combination with the α/β-hydrolase fold receptor (D14 or DAD2) catalyzes the ubiquitination of transcriptional repressors, such as D53 and SMXL6/7/8, upon binding of the SL ligand to the receptor; this results in the degradation of these repressors by the 26S proteasome, and subsequent activation of the transcription of SL responsive genes. The SL receptor differs from receptors of other plant hormones, as it acts as an enzyme that hydrolyses the SL ligand and binds covalently to the released D-ring moiety.

It can be assumed that the structural diversity of SLs is a result of an evolution toward particular functions in the communication with other organisms and in regulating plant development and stress response. Hence, the identification of this functional specificity and the elucidation of the biosynthesis routes leading to the different SLs are expected to open up new possibilities in developing crops with optimized architecture, increased resistance to root parasitic weeds, and/or higher mycorrhization rates. Knowledge about particular functions of different SLs may also pave the way for designing SL analogs with specific applications.

1.7 Application Potential for SLs

The diversity of the functions predestines SLs, SL analogs/mimics, and antagonists for application in agriculture and horticulture and makes the biosynthesis, release, and perception of this hormone an excellent target for genetic engineering and breeding toward crops with improved performance.

Optimal plant architecture is essential for exploiting the production and yield potential of crops. SLs are a key regulator of shoot and root architecture, as suggested by the effect of SL deficiency on shoot branching and plant height. Therefore, modulation of SL biosynthesis and perception holds promise for generating crops and ornamental plants with the architecture, e.g., tiller/branch number, pattern of roots types, etc., desired for different agriculture and horticulture systems. Moreover, the recently discovered role of SLs in plant response to biotic and abiotic stress opens up a further possibility of enhancing crop productivity by manipulating SL biosynthesis and response toward increasing drought, salt, and/or pathogen resistance. In this context, application of SL analogs and antagonists might be a promising, more straightforward approach to improve the resilience and architecture of plants in agriculture and horticulture.

The root parasitic weed *S. hermonthica* is a major threat to global food security, affecting many regions in the world, particularly sub-Saharan Africa. SLs are a major factor in the Striga/host plant interplay and a key component in dealing with this threat. Recent data indicate that pattern and amount of released SLs determine the susceptibility to Striga. Strigol-type SLs seem to be much better germination stimulants, compared to the orobanchol-like ones. Related *Phelipanche* and *Orobanche* spp. have also adapted to certain type of SLs. Hence, alteration of SLs pattern is an attractive possibility to alleviate the infestation by root parasitic weeds.

This alteration can be approached either by breeding or by genetic engineering. Better knowledge on late steps of SL biosynthesis as well as on release mechanisms, which determine type and of released SLs, is required to achieve this target.

Considering the essential role of SLs in establishing the mycorrhizal symbiosis and the different efficiency of various SLs in this process, understanding the biological reason for SL diversity and the mechanism of their release will also help in increasing the mycorrhization, which improves the growth of host plants and reduce the demand for fertilizers.

Root parasitic plants produce very large amounts of seeds that can remain viable for more than a decade in the soil. Thus, seed banks of Striga and other root parasitic plants, which have accumulated over decades in infested soils, are major constraints in combating these weeds. The application of SL analogs to induce seed germination of these obligate parasites in the absence of a host is a promising approach that can significantly reduce or even eliminate accumulated seed bank. This method, suicidal germination, can be combined with other approaches, such as planting varieties with increased resistance, in an integrated strategy to successfully control root parasitic weeds.

Acknowledgment We thank Justine Braguy and Jianing Mi for assisting drawing the illustrations and their critical reading to the manuscript.

Glossary

ABC transporter ATP-binding cassette transporters, consisting of multi-subunits including transmembrane proteins and membrane-associated ATPases, with especially important roles in transport of plant secondary metabolites and hormones.

α/β-fold Hydrolase A large, diverse superfamily of hydrolytic enzymes character-ized by a core alpha-/beta-sheet, which contains eight beta strands connected by six alpha helices and a catalytic triad.

Apocarotenoids The oxidative cleavage products of carotenoids by CCDs or spontaneous oxidation.

Arbuscular mycorrhizal (AM) fungi A class of symbiotic fungi of the phylum *Glomeromycota*, characterized by the formation of unique intracellular structures called arbuscules that receive organic carbon from the host and assist the plant in the acquisition of mineral nutrients through their associations with roots.

Butenolide A lactone with a four-carbon heterocyclic ring structure. It is a common moiety in all SLs.

Canonical SLs A subfamily of SLs characterized by the presence of a tricyclic lactone (ABC-ring) connected to a conserved butenolide ring (D-ring) via an enol ether bridge in *R*-configuration.

Carlactone A core intermediate in the biosynthesis of SLs, generated by the sequential action of D27, CCD7, and CCD8 from all-*trans*-β-carotene.

Carotenoids A class of terpenoid pigments produced in plants, algae, and some bacteria. They fulfill essential functions in photosynthesis and serve as precursors of hormones and signaling molecules.

Carotenoid cleavage dioxygenases (CCDs) A large family of non-heme iron (II)-dependent enzymes which break C=C double bonds in carotenoid or apocarotenoid backbone, leading to two carbonyl products.

Catalytic triad A set of three coordinated amino acids comprising an acid, a base, and a nucleophile (often Asp, His, and Ser, respectively) found in the active site of hydrolases.

F-box protein A component of the SCF-type E3 ubiquitin-protein ligase complexes, which are responsible for substrate recognition, polyubiquitination, and eventually protein degradation.

HPLC High-performance liquid chromatography, an analytical chemistry technique used to separate, identify, and quantify different compounds in a sample mixture, which relies on pumps to pass a pressurized liquid solvent containing the sample mixture through a column filled with a solid adsorbent material. Due to the slightly different interaction of each substance in the sample with the adsorbent material, different substances have different flow rates when flowing out of the column therefore leading to their separation.

LC-MS Liquid chromatography-mass spectrometry, a commonly used technique in analytical chemistry to identify a chemical by combining liquid chromatography (LC) or high-performance liquid chromatography (HPLC) with the mass analysis capabilities of mass spectrometry (MS).

MEP pathway 2-*C*-Methyl-D-erythritol 4-phosphate pathway, a route for the biosynthesis of the isoprenoid precursor isopentenyl pyrophosphate (IPP), which starts with the condensation of pyruvate with D-glyceraldehyde phosphate. The MEP pathway is responsible for the synthesis of the isoprenoid building block IPP in bacteria and plastids.

Mevalonate pathway A pathway for the synthesis of isopentenyl pyrophosphate (IPP) in the cytoplasm of eukaryotic cells, archaea, and some bacteria. The mevalonate pathway is initiated by the condensation of two molecules acetyl-CoA and is the source of IPP in the cytoplasm of eukaryotic cells.

MS Mass spectrometry, an analytical technique that ionizes chemical species by electrons, ions or photons, energetic neutral atoms, or heavy cluster ions and sorts the ions based on their mass-to-charge ratio (m/z) and to detect them qualitatively and quantitatively by their respective m/z and abundance.

Rhizosphere The region of soil surrounding the roots, which is directly affected by root secretions and is enriched in soil microorganisms.

Rootstock The lower part of the combined grafted plant.

Scion The upper part of the combined grafted plant.

Sesquiterpenes A class of terpenes formed by the condensation of three isoprene units and consisting of a C_{15} skeleton.

Non-canonical SLs Subfamily of SLs that contain a variable second moiety instead of the tricyclic lactone connected to a conserved butenolide ring (D-ring) via an enol ether bridge in *R*-configuration.

References

Abe S, Sado A, Tanaka K, Kisugi T, Asami K, Ota S, Kim HI, Yoneyama K, Xie X, Ohnishi T (2014) Carlactone is converted to carlactonoic acid by MAX1 in Arabidopsis and its methyl ester can directly interact with AtD14 in vitro. Proc Natl Acad Sci USA 111:18084–18089

Abuauf H, Haider I, Jia K-P, Ablazov A, Mi J, Blilou I, Al-Babili S (2018) The Arabidopsis DWARF27 gene encodes an *all-trans-/9-cis*-β-carotene isomerase and is induced by auxin, abscisic acid and phosphate deficiency. Plant Sci 277:33–42

Aguilar-Martinez JA, Poza-Carrion C, Cubas P (2007) Arabidopsis BRANCHED1 acts as an integrator of branching signals within axillary buds. Plant Cell 19:458–472

Akiyama K, Matsuzaki K-I, Hayashi H (2005) Plant sesquiterpenes induce hyphal branching in arbuscular mycorrhizal fungi. Nature 435:824–827

Al-Babili S, Bouwmeester HJ (2015) Strigolactones, a novel carotenoid-derived plant hormone. Annu Rev Plant Biol 66:161–186

Alder A, Holdermann I, Beyer P, Al-Babili S (2008) Carotenoid oxygenases involved in plant branching catalyse a highly specific conserved apocarotenoid cleavage reaction. Biochem J 416:289–296

Alder A, Jamil M, Marzorati M, Bruno M, Vermathen M, Bigler P, Ghisla S, Bouwmeester H, Beyer P, Al-Babili S (2012) The path from beta-carotene to carlactone, a strigolactone-like plant hormone. Science 335:1348–1351

Arite T, Iwata H, Ohshima K, Maekawa M, Nakajima M, Kojima M, Sakakibara H, Kyozuka J (2007) DWARF10, an RMS1/MAX4/DAD1 ortholog, controls lateral bud outgrowth in rice. Plant J 51:1019–1029

Arite T, Umehara M, Ishikawa S, Hanada A, Maekawa M, Yamaguchi S, Kyozuka J (2009) d14, a strigolactone-insensitive mutant of rice, shows an accelerated outgrowth of tillers. Plant Cell Physiol 50:1416–1424

Auldridge ME, McCarty DR, Klee HJ (2006) Plant carotenoid cleavage oxygenases and their apocarotenoid products. Curr Opin Plant Biol 9:315–321

Avendano-Vazquez AO, Cordoba E, Llamas E, San Roman C, Nisar N, De la Torre S, Ramos-Vega M, Gutierrez-Nava MD, Cazzonelli CI, Pogson BJ, Leon P (2014) An uncharacterized apocarotenoid-derived signal generated in zeta-carotene desaturase mutants regulates leaf development and the expression of chloroplast and nuclear genes in arabidopsis. Plant Cell 26 (6):2524–2537

Awad AA, Sato D, Kusumoto D, Kamioka H, Takeuchi Y, Yoneyama K (2006) Characterization of strigolactones, germination stimulants for the root parasitic plants Striga and Orobanche, produced by maize, millet and sorghum. Plant Growth Regul 48:221

Baz L, Mori N, Guo X, Jamil M, Kountche BA, Mi J, Jia K-P, Vermathen M, Akiyama K, Al-Babili S (2018) 3-Hydroxycarlactone, a novel product of the strigolactone biosynthesis core pathway. Mol Plant 11(10):1312–1314

Beveridge CA, Ross JJ, Murfet IC (1996) Branching in pea (action of genes Rms3 and Rms4). Plant Physiol 110:859–865

Bonfante P, Genre A (2015) Arbuscular mycorrhizal dialogues: do you speak 'plantish' or 'fungish'? Trends Plant Sci 20:150–154

Bonneau L, Huguet S, Wipf D, Pauly N, Truong HN (2013) Combined phosphate and nitrogen limitation generates a nutrient stress transcriptome favorable for arbuscular mycorrhizal symbiosis in *Medicago truncatula*. New Phytol 199:188–202

Booker J, Auldridge M, Wills S, McCarty D, Klee H, Leyser O (2004) MAX3/CCD7 is a carotenoid cleavage dioxygenase required for the synthesis of a novel plant signaling molecule. Curr Biol 14:1232–1238

Booker J, Sieberer T, Wright W, Williamson L, Willett B, Stirnberg P, Turnbull C, Srinivasan M, Goddard P, Leyser O (2005) MAX1 encodes a cytochrome P450 family member that acts downstream of MAX3/4 to produce a carotenoid-derived branch-inhibiting hormone. Dev Cell 8:443–449

Bouwmeester HJ, Matusova R, Zhongkui S, Beale MH (2003) Secondary metabolite signalling in host–parasitic plant interactions. Curr Opin Plant Biol 6:358–364

Braun N, de Saint Germain A, Pillot J-P, Boutet-Mercey S, Dalmais M, Antoniadi I, Li X, Maia-Grondard A, Le Signor C, Bouteiller N (2012) The pea TCP transcription factor PsBRC1 acts downstream of strigolactones to control shoot branching. Plant Physiol 158:225–238

Breuillin F, Schramm J, Hajirezaei M, Ahkami A, Favre P, Druege U, Hause B, Bucher M, Kretzschmar T, Bossolini E (2010) Phosphate systemically inhibits development of arbuscular mycorrhiza in Petunia hybrida and represses genes involved in mycorrhizal functioning. Plant J 64:1002–1017

Brewer PB, Koltai H, Beveridge CA (2013) Diverse roles of strigolactones in plant development. Mol Plant 6:18–28

Brewer PB, Yoneyama K, Filardo F, Meyers E, Scaffidi A, Frickey T, Akiyama K, Seto Y, Dun EA, Cremer JE, Kerr SC, Waters MT, Flematti GR, Mason MG, Weiller G, Yamaguchi S, Nomura T, Smith SM, Yoneyama K, Beveridge CA (2016) LATERAL BRANCHING OXI-DOREDUCTASE acts in the final stages of strigolactone biosynthesis in Arabidopsis. Proc Natl Acad Sci USA 113:6301–6306

Britton G (1995) Structure and properties of carotenoids in relation to function. FASEB J 9:1551–1558

Bruno M, Al-Babili S (2016) On the substrate specificity of the rice strigolactone biosynthesis enzyme DWARF27. Planta 243:1429–1440

Bruno M, Hofmann M, Vermathen M, Alder A, Beyer P, Al-Babili S (2014) On the substrate-and stereospecificity of the plant carotenoid cleavage dioxygenase 7. FEBS Lett 588:1802–1807

Bruno M, Beyer P, Al-Babili S (2015) The potato carotenoid cleavage dioxygenase 4 catalyzes a single cleavage of β-ionone ring-containing carotenes and non-epoxidated xanthophylls. Arch Biochem Biophys 572:126–133

Bruno M, Koschmieder J, Wuest F, Schaub P, Fehling-Kaschek M, Timmer J, Beyer P, Al-Babili S (2016) Enzymatic study on AtCCD4 and AtCCD7 and their potential to form acyclic regulatory metabolites. J Exp Bot 67:5993–6005

Bruno M, Vermathen M, Alder A, Wüst F, Schaub P, Steen R, Beyer P, Ghisla S, Al-Babili S (2017) Insights into the formation of carlactone from in-depth analysis of the CCD8-catalyzed reactions. FEBS Lett 591:792–800

Butler LG (1995) Chemical communication between the parasitic weed striga and its crop host. ACS Symp Ser 582:158–168

Campbell R, Ducreux LJ, Morris WL, Morris JA, Suttle JC, Ramsay G, Bryan GJ, Hedley PE, Taylor MA (2010) The metabolic and developmental roles of carotenoid cleavage dioxygenase4 from potato. Plant Physiol 154:656–664

Challis RJ, Hepworth J, Mouchel C, Waites R, Leyser O (2013) A role for more axillary growth1 (MAX1) in evolutionary diversity in strigolactone signaling upstream of MAX2. Plant Physiol 161:1885–1902

Chapple C (1998) Molecular-genetic analysis of plant cytochrome P450-dependent monooxygenases. Annu Rev Plant Biol 49:311–343

Charnikhova TV, Gaus K, Lumbroso A, Sanders M, Vincken J-P, De Mesmaeker A, Ruyter-Spira CP, Screpanti C, Bouwmeester HJ (2017) Zealactones. Novel natural strigolactones from maize. Phytochemistry 137:123–131

Charnikhova TV, Gaus K, Lumbroso A, Sanders M, Vincken J-P, De Mesmaeker A, Ruyter-Spira CP, Screpanti C, Bouwmeester HJ (2018) Zeapyranolactone − a novel strigolactone from maize. Phytochem Lett 24:172–178

Chhikara N, Kour R, Jaglan S, Gupta P, Gat Y, Panghal A (2018) *Citrus medica*: nutritional, phytochemical composition and health benefits–a review. Food Funct 9:1978–1992

Conn CE, Bythell-Douglas R, Neumann D, Yoshida S, Whittington B, Westwood JH, Shirasu K, Bond CS, Dyer KA, Nelson DC (2015) Convergent evolution of strigolactone perception enabled host detection in parasitic plants. Science 349:540–543

Cook CE, Whichard LP, Turner B, Wall ME, Egley GH (1966) Germination of Witchweed (Striga lutea Lour.): isolation and properties of a potent stimulant. Science 154:1189–1190

Cook C, Whichard LP, Wall M, Egley GH, Coggon P, Luhan PA, McPhail A (1972) Germination stimulants. II. Structure of strigol, a potent seed germination stimulant for witchweed (*Striga lutea*). J Am Chem Soc 94:6198–6199

Cramer WA, Zhang H, Yan J, Kurisu G, Smith JL (2006) Transmembrane traffic in the cytochrome b 6 f complex. Annu Rev Biochem 75:769–790

de Kraker J-W, Franssen MC, de Groot A, König WA, Bouwmeester HJ (1998) (+)-Germacrene A biosynthesis: the committed step in the biosynthesis of bitter sesquiterpene lactones in chicory. Plant Physiol 117:1381–1392

de Saint Germain A, Clavé G, Badet-Denisot M-A, Pillot J-P, Cornu D, Le Caer J-P, Burger M, Pelissier F, Retailleau P, Turnbull C (2016) An histidine covalent receptor and butenolide complex mediates strigolactone perception. Nat Chem Biol 12:787–794

Decker EL, Alder A, Hunn S, Ferguson J, Lehtonen MT, Scheler B, Kerres KL, Wiedemann G, Safavi-Rizi V, Nordzieke S (2017) Strigolactone biosynthesis is evolutionarily conserved, regulated by phosphate starvation and contributes to resistance against phytopathogenic fungi in a moss, *Physcomitrella patens*. New Phytol 216(2):455–468

Delavault P, Simier P, Thoiron S, Véronési C, Fer A, Thalouarn P (2002) Isolation of mannose 6-phosphate reductase cDNA, changes in enzyme activity and mannitol content in broomrape (*Orobanche ramosa*) parasitic on tomato roots. Physiol Plant 115:48–55

DellaPenna D, Pogson BJ (2006) Vitamin synthesis in plants: tocopherols and carotenoids. Annu Rev Plant Biol 57:711–738

Dharmasiri N, Dharmasiri S, Estelle M (2005) The F-box protein TIR1 is an auxin receptor. Nature 26(435(7041)):441–445

Doebley J, Stec A, Hubbard L (1997) The evolution of apical dominance in maize. Nature 386 (6624):485–488

Drummond RS, Martínez-Sánchez NM, Janssen BJ, Templeton KR, Simons JL, Quinn BD, Karunairetnam S, Snowden KC (2009) *Petunia hybrida* CAROTENOID CLEAVAGE DIOXYGENASE7 is involved in the production of negative and positive branching signals in petunia. Plant Physiol 151:1867–1877

Drummond RS, Sheehan H, Simons JL, Martínez-Sánchez NM, Turner RM, Putterill J, Snowden KC (2011) The expression of petunia strigolactone pathway genes is altered as part of the endogenous developmental program. Front Plant Sci 10(2):115

Estrada AF, Maier D, Scherzinger D, Avalos J, Al-Babili S (2008) Novel apocarotenoid intermediates in Neurospora crassa mutants imply a new biosynthetic reaction sequence leading to neurosporaxanthin formation. Fungal Genet Biol 45:1497–1505

Flematti GR, Dixon KW, Smith SM (2015) What are karrikins and how were they 'discovered' by plants? BMC Biol 13:108

Floss DS, Schliemann W, Schmidt J, Strack D, Walter MH (2008) RNA interference-mediated repression of *MtCCD1* in mycorrhizal roots of *Medicago truncatula* causes accumulation of C27 apocarotenoids, shedding light on the functional role of CCD1. Plant Physiol 148:1267–1282

Foo E, Bullier E, Goussot M, Foucher F, Rameau C, Beveridge CA (2005) The branching gene RAMOSUS1 mediates interactions among two novel signals and auxin in pea. Plant Cell 17:464–474

Fraser PD, Bramley PM (2004) The biosynthesis and nutritional uses of carotenoids. Prog Lipid Res 43:228–265

Fukui K, Ito S, Ueno K, Yamaguchi S, Kyozuka J, Asami T (2011) New branching inhibitors and their potential as strigolactone mimics in rice. Bioorg Med Chem Lett 21:4905–4908

Giuliano G, Al-Babili S, Von Lintig J (2003) Carotenoid oxygenases: cleave it or leave it. Trends Plant Sci 8:145–149

Gobena D, Shimels M, Rich PJ, Ruyter-Spira C, Bouwmeester H, Kanuganti S, Mengiste T, Ejeta G (2017) Mutation in sorghum LOW GERMINATION STIMULANT 1 alters strigolactones and causes Striga resistance. Proc Natl Acad Sci USA 114:4471–4476

Gomez-Roldan V, Fermas S, Brewer PB, Puech-Pages V, Dun EA, Pillot JP, Letisse F, Matusova R, Danoun S, Portais JC, Bouwmeester H, Becard G, Beveridge CA, Rameau C, Rochange SF (2008) Strigolactone inhibition of shoot branching. Nature 455:189–194

Gonzalez-Jorge S, Ha S-H, Magallanes-Lundback M, Gilliland LU, Zhou A, Lipka AE, Nguyen Y-N, Angelovici R, Lin H, Cepela J (2013) Carotenoid cleavage dioxygenase4 is a negative regulator of β-carotene content in Arabidopsis seeds. Plant Cell 25:4812–4826

Goodwin TW (1988) Plant pigments. Academic Press, London

Guan JC, Koch KE, Suzuki M, Wu S, Latshaw S, Petruff T, Goulet C, Klee HJ, McCarty DR (2012) Diverse roles of strigolactone signaling in maize architecture and the uncoupling of a branching-specific subnetwork. Plant Physiol 160:1303–1317

Guillotin B, Etemadi M, Audran C, Bouzayen M, Bécard G, Combier JP (2017) Sl-IAA27 regulates strigolactone biosynthesis and mycorrhization in tomato (var. MicroTom). New Phytol 213:1124–1132

Guo Y, Zheng Z, La Clair JJ, Chory J, Noel JP (2013) Smoke-derived karrikin perception by the α/β-hydrolase KAI2 from Arabidopsis. Proc Natl Acad Sci USA 110:8284–8289

Gutjahr C (2014) Phytohormone signaling in arbuscular mycorrhiza development. Curr Opin Plant Biol 20:26–34

Gutjahr C, Parniske M (2013) Cell and developmental biology of arbuscular mycorrhiza symbiosis. Annu Rev Cell Dev Biol 29:593–617

Haider I, Andreo-Jimenez B, Bruno M, Bimbo A, Floková K, Abuauf H, Ntui VO, Guo X, Charnikhova T, Al-Babili S (2018) The interaction of strigolactones with abscisic acid during the drought response in rice. J Exp Bot 69:2403–2414

Hamiaux C, Drummond RS, Janssen BJ, Ledger SE, Cooney JM, Newcomb RD, Snowden KC (2012) DAD2 is an α/β hydrolase likely to be involved in the perception of the plant branching hormone, strigolactone. Curr Biol 22:2032–2036

Hashimot H, Uragami C, Cogdell RJ (2016) Carotenoids and photosynthesis. In: Stange C (ed) Carotenoids in nature, Subcellular biochemistry, vol 79. Springer, Cham

Hayward A, Stirnberg P, Beveridge C, Leyser O (2009) Interactions between auxin and strigolactone in shoot branching control. Plant Physiol 151:400–412

Hershko A, Ciechanover A (1998) The ubiquitin system. Ann Rev 67:425–479

Howitt CA, Pogson BJ (2006) Carotenoid accumulation and function in seeds and non-green tissues. Plant Cell Environ 29:435–445

Huang F-C, Horváth G, Molnár P, Turcsi E, Deli J, Schrader J, Sandmann G, Schmidt H, Schwab W (2009) Substrate promiscuity of RdCCD1, a carotenoid cleavage oxygenase from Rosa damascena. Phytochemistry 70:457–464

Ilg A, Yu Q, Schaub P, Beyer P, Al-Babili S (2010) Overexpression of the rice carotenoid cleavage dioxygenase 1 gene in Golden Rice endosperm suggests apocarotenoids as substrates in planta. Planta 232:691–699

Iseki M, Shida K, Kuwabara K, Wakabayashi T, Mizutani M, Takikawa H, Sugimoto Y (2018) Evidence for species-dependent biosynthetic pathways for converting carlactone to strigolactones in plants. J Exp Bot 69:2305–2318

Ishikawa S, Maekawa M, Arite T, Onishi K, Takamure I, Kyozuka J (2005) Suppression of tiller bud activity in tillering dwarf mutants of rice. Plant Cell Physiol 46:79–86

Isin EM, Guengerich FP (2007) Complex reactions catalyzed by cytochrome P450 enzymes. Biochim Biophys Acta Gen Subj 1770:314–329

Ito S, Yamagami D, Umehara M, Hanada A, Yoshida S, Sasaki Y, Yajima S, Kyozuka J, Ueguchi-Tanaka M, Matsuoka M (2017) Regulation of strigolactone biosynthesis by gibberellin signaling. Plant Physiol 174:1250–1259

Jamil M, Rodenburg J, Charnikhova T, Bouwmeester HJ (2011) Pre-attachment *Striga hermonthica* resistance of New Rice for Africa (NERICA) cultivars based on low strigolactone production. New Phytol 192:964–975

Jamil M, Kanampiu F, Karaya H, Charnikhova T, Bouwmeester H (2012) *Striga hermonthica* parasitism in maize in response to N and P fertilisers. Field Crop Res 134:1–10

Jamil M, Van Mourik T, Charnikhova T, Bouwmeester H (2013) Effect of diammonium phosphate application on strigolactone production and *Striga hermonthica* infection in three sorghum cultivars. Weed Res 53:121–130

Jamil M, Kountche BA, Haider I, Guo X, Ntui VO, Jia K-P, Ali S, Hameed US, Nakamura H, Lyu Y (2017) Methyl phenlactonoates are efficient strigolactone analogs with simple structure. J Exp Bot 69(9):2319–2331

Jia KP, Kountche BA, Jamil M, Guo X, Ntui VO, Rüfenacht A, Rochange S, Al-Babili S (2016) Nitro-phenlactone, a carlactone analog with pleiotropic strigolactone activities. Mol Plant 9:1341–1344

Jia KP, Baz L, Al-Babili S (2018) From carotenoids to strigolactones. J Exp Bot 69(9):2189–2204

Jiang L, Liu X, Xiong G, Liu H, Chen F, Wang L, Meng X, Liu G, Yu H, Yuan Y (2013) DWARF 53 acts as a repressor of strigolactone signalling in rice. Nature 504:401

Jiao Y, Wang Y, Xue D, Wang J, Yan M, Liu G, Dong G, Zeng D, Lu Z, Zhu X (2010) Regulation of OsSPL14 by OsmiR156 defines ideal plant architecture in rice. Nat Genet 42:541

Johnson X, Brcich T, Dun EA, Goussot M, Haurogné K, Beveridge CA, Rameau C (2006) Branching genes are conserved across species. Genes controlling a novel signal in pea are coregulated by other long-distance signals. Plant Physiol 142:1014–1026

Kagiyama M, Hirano Y, Mori T, Kim SY, Kyozuka J, Seto Y, Yamaguchi S, Hakoshima T (2013) Structures of D14 and D14L in the strigolactone and karrikin signaling pathways. Genes Cells 18:147–160

Katsir L, Chung HS, Koo AJ, Howe GA (2008) Jasmonate signaling: a conserved mechanism of hormone sensing. Curr Opin Plant Biol 11:428–435

Ke J, Ma H, Gu X, Thelen A, Brunzelle JS, Li J, Xu HE, Melcher K (2015) Structural basis for recognition of diverse transcriptional repressors by the TOPLESS family of corepressors. Sci Adv 1:e1500107

Kepinski S, Leyser O (2005) The Arabidopsis F-box protein TIR1 is an auxin receptor. Nature 435:446

Khosla A, Nelson DC (2016) Strigolactones, super hormones in the fight against Striga. Curr Opin Plant Biol 33:57–63

Kim HI, Kisugi T, Khetkam P, Xie X, Yoneyama K, Uchida K, Yokota T, Nomura T, McErlean CS, Yoneyama K (2014) Avenaol, a germination stimulant for root parasitic plants from *Avena strigosa*. Phytochemistry 103:85–88

Kohlen W, Charnikhova T, Liu Q, Bours R, Domagalska MA, Beguerie S, Verstappen F, Leyser O, Bouwmeester HJ, Ruyter-Spira C (2011) Strigolactones are transported through the xylem and play a key role in shoot architectural response to phosphate deficiency in non-AM host *Arabidopsis thaliana*. Plant Physiol 110:164640

Kohlen W, Charnikhova T, Bours R, López-Ráez JA, Bouwmeester H (2013) Tomato strigolactones: a more detailed look. Proc Natl Acad Sci USA 8:e22785

Koltai H (2011) Strigolactones are regulators of root development. New Phytol 190:545–549

Kretzschmar T, Kohlen W, Sasse J, Borghi L, Schlegel M, Bachelier JB, Reinhardt D, Bours R, Bouwmeester HJ, Martinoia E (2012) A petunia ABC protein controls strigolactone-dependent symbiotic signalling and branching. Nature 483:341–344

Lakshminarayana R, Raju M, Krishnakantha TP, Baskaran V (2005) Determination of major carotenoids in a few Indian leafy vegetables by high-performance liquid chromatography. J Agric Food Chem 53:2838–2842

Lin H, Wang R, Qian Q, Yan M, Meng X, Fu Z, Yan C, Jiang B, Su Z, Li J (2009) DWARF27, an iron-containing protein required for the biosynthesis of strigolactones, regulates rice tiller bud outgrowth. Plant Cell 21:1512–1525

Liu J, He H, Vitali M, Visentin I, Charnikhova T, Haider I, Schubert A, Ruyter-Spira C, Bouwmeester HJ, Lovisolo C (2015) Osmotic stress represses strigolactone biosynthesis in Lotus japonicus roots: exploring the interaction between strigolactones and ABA under abiotic stress. Planta 241:1435–1451

Lopez-Raez JA, Kohlen W, Charnikhova T, Mulder P, Undas AK, Sergeant MJ, Verstappen F, Bugg TD, Thompson AJ, Ruyter-Spira C, Bouwmeester H (2010) Does abscisic acid affect strigolactone biosynthesis? New Phytol 187:343–354

Ma G, Zhang L, Matsuta A, Matsutani K, Yamawaki K, Yahata M, Wahyudi A, Motohashi R, Kato M (2013) Enzymatic formation of β-citraurin from β-cryptoxanthin and zeaxanthin by carotenoid cleavage dioxygenase4 in the flavedo of citrus fruit. Plant Physiol 163:682–695

Maass D, Arango J, Wüst F, Beyer P, Welsch R (2009) Carotenoid crystal formation in Arabidopsis and carrot roots caused by increased phytoene synthase protein levels. PLoS One 4:e6373

Mashiguchi K, Sasaki E, Shimada Y, Nagae M, Ueno K, Nakano T, Yoneyama K, Suzuki Y, Asami T (2009) Feedback-regulation of strigolactone biosynthetic genes and strigolactone-regulated genes in Arabidopsis. Biosci Biotechnol Biochem 73:2460–2465

Matusova R, Rani K, Verstappen FW, Franssen MC, Beale MH, Bouwmeester HJ (2005) The strigolactone germination stimulants of the plant-parasitic Striga and Orobanche spp. are derived from the carotenoid pathway. Plant Physiol 139:920–934

Medina HR, Cerdá-Olmedo E, Al-Babili S (2011) Cleavage oxygenases for the biosynthesis of trisporoids and other apocarotenoids in Phycomyces. Mol Microbiol 82:199–208

Minakuchi K, Kameoka H, Yasuno N, Umehara M, Luo L, Kobayashi K, Hanada A, Ueno K, Asami T, Yamaguchi S (2010) FINE CULM1 (FC1) works downstream of strigolactones to inhibit the outgrowth of axillary buds in rice. Plant Cell Physiol 51:1127–1135

Miura K, Ikeda M, Matsubara A, Song X-J, Ito M, Asano K, Matsuoka M, Kitano H, Ashikari M (2010) OsSPL14 promotes panicle branching and higher grain productivity in rice. Nat Genet 42:545

Mockaitis K, Estelle M (2008) Auxin receptors and plant development: a new signaling paradigm. Annu Rev Cell Dev Biol 24:55–80

Moise AR, Von Lintig J, Palczewski K (2005) Related enzymes solve evolutionarily recurrent problems in the metabolism of carotenoids. Trends Plant Sci 10:178–186

Moise AR, Al-Babili S, Wurtzel ET (2014) Mechanistic aspects of carotenoid biosynthesis. Chem Rev 114:164–193

Morris SE, Turnbull CG, Murfet IC, Beveridge CA (2001) Mutational analysis of branching in pea. Evidence ThatRms1 and Rms5 regulate the same novel signal. Plant Physiol 126:1205–1213

Motonami N, Ueno K, Nakashima H, Nomura S, Mizutani M, Takikawa H, Sugimoto Y (2013) The bioconversion of 5-deoxystrigol to sorgomol by the sorghum, Sorghum bicolor (L.) Moench. Phytochemistry 93:41–48

Nakamura H, Xue Y-L, Miyakawa T, Hou F, Qin H-M, Fukui K, Shi X, Ito E, Ito S, Park S-H (2013) Molecular mechanism of strigolactone perception by DWARF14. Nat Commun 4:2613

Nelson DC, Riseborough JA, Flematti GR, Stevens J, Ghisalberti EL, Dixon KW, Smith SM (2009) Karrikins discovered in smoke trigger Arabidopsis seed germination by a mechanism requiring gibberellic acid synthesis and light. Plant Physiol 149:863–873

Nelson DC, Flematti GR, Riseborough JA, Ghisalberti EL, Dixon KW, Smith SM (2010) Karrikins enhance light responses during germination and seedling development in Arabidopsis thaliana. Proc Natl Acad Sci USA 107:7095–7100

Nisar N, Li L, Lu S, Khin NC, Pogson BJ (2015) Carotenoid metabolism in plants. Mol Plant 8:68–82

Ohmiya A, Kishimoto S, Aida R, Yoshioka S, Sumitomo K (2006) Carotenoid cleavage dioxygenase (CmCCD4a) contributes to white color formation in chrysanthemum petals. Plant Physiol 142:1193–1201

Parker C (2009) Observations on the current status of Orobanche and Striga problems worldwide. Pest Manag Sci 65:453–459

Parry AD, Horgan R (1992) Abscisic acid biosynthesis in roots. Planta 187:185–191

Pauwels L, Barbero GF, Geerinck J, Tilleman S, Grunewald W, Pérez AC, Chico JM, Bossche RV, Sewell J, Gil E (2010) NINJA connects the co-repressor TOPLESS to jasmonate signalling. Nature 464:788

Rodrigo MJ, Alquézar B, Alós E, Medina V, Carmona L, Bruno M, Al-Babili S, Zacarías L (2013) A novel carotenoid cleavage activity involved in the biosynthesis of Citrus fruit-specific apocarotenoid pigments. J Exp Bot 64:4461–4478

Ruiz-Sola MÁ, Rodríguez-Concepción M (2012) Carotenoid biosynthesis in Arabidopsis: a colorful pathway. Arabidopsis Book 10:e0158

Ruyter-Spira C, Al-Babili S, Van Der Krol S, Bouwmeester H (2013) The biology of strigolactones. Trends Plant Sci 18:72–83

Sah SK, Reddy KR, Li J (2016) Abscisic acid and abiotic stress tolerance in crop plants. Front Plant Sci 4(7):571

Samodelov SL, Beyer HM, Guo X, Augustin M, Jia K-P, Baz L, Ebenhöh O, Beyer P, Weber W, Al-Babili S (2016) StrigoQuant: a genetically encoded biosensor for quantifying strigolactone activity and specificity. Sci Adv 2.:e1601266:1–8

Santner A, Estelle M (2009) Recent advances and emerging trends in plant hormone signalling. Nature 459:1071

Sasse J, Simon S, Gubeli C, Liu GW, Cheng X, Friml J, Bouwmeester H, Martinoia E, Borghi L (2015) Asymmetric localizations of the ABC transporter PaPDR1 trace paths of directional strigolactone transport. Curr Biol 25:647–655

Scaffidi A, Waters MT, Sun YK, Skelton BW, Dixon KW, Ghisalberti EL, Flematti GR, Smith SM (2014) Strigolactone hormones and their stereoisomers signal through two related receptor proteins to induce different physiological responses in Arabidopsis. Plant Physiol 165:1221–1232

Scherzinger D, Al-Babili S (2008) In vitro characterization of a carotenoid cleavage dioxygenase from Nostoc sp. PCC 7120 reveals a novel cleavage pattern, cytosolic localization and induction by highlight. Mol Microbiol 69:231–244

Schlicht M, Šamajová O, Schachtschabel D, Mancuso S, Menzel D, Boland W, Baluška F (2008) D'orenone blocks polarized tip growth of root hairs by interfering with the PIN2-mediated auxin transport network in the root apex. Plant J 55:709–717

Schwartz SH, Tan BC, Gage DA, Zeevaart JA, McCarty DR (1997) Specific oxidative cleavage of carotenoids by VP14 of maize. Science 276:1872–1874

Schwartz SH, Qin X, Loewen MC (2004) The biochemical characterization of two carotenoid cleavage enzymes from Arabidopsis indicates that a carotenoid-derived compound inhibits lateral branching. J Biol Chem 279:46940–46945

Seto Y, Sado A, Asami K, Hanada A, Umehara M, Akiyama K, Yamaguchi S (2014) Carlactone is an endogenous biosynthetic precursor for strigolactones. Proc Natl Acad Sci USA 111:1640–1645

Shen H, Zhu L, Bu QY, Huq E (2012) MAX2 affects multiple hormones to promote photomorphogenesis. Mol Plant 5:750–762

Shu K, Yang W (2017) E3 ubiquitin ligases: ubiquitous actors in plant development and abiotic stress responses. Plant Cell Physiol 58:1461–1476

Siame BA, Weerasuriya Y, Wood K, Ejeta G, Butler LG (1993) Isolation of strigol, a germination stimulant for Striga asiatica, from host plants. J Agric Food Chem 41:1486–1491

Simons JL, Napoli CA, Janssen BJ, Plummer KM, Snowden KC (2007) Analysis of the DECREASED APICAL DOMINANCE genes of petunia in the control of axillary branching. Plant Physiol 143:697–706

Snowden KC, Simkin AJ, Janssen BJ, Templeton KR, Loucas HM, Simons JL, Karunairetnam S, Gleave AP, Clark DG, Klee HJ (2005) The decreased apical dominance1/Petunia hybrida CAROTENOID CLEAVAGE DIOXYGENASE8 gene affects branch production and plays a role in leaf senescence, root growth, and flower development. Plant Cell 17:746–759

Song X, Lu Z, Yu H, Shao G, Xiong J, Meng X, Jing Y, Liu G, Xiong G, Duan J (2017) IPA1 functions as a downstream transcription factor repressed by D53 in strigolactone signaling in rice. Cell Res 27:1128

Sorefan K, Booker J, Haurogné K, Goussot M, Bainbridge K, Foo E, Chatfield S, Ward S, Beveridge C, Rameau C (2003) MAX4 and RMS1 are orthologous dioxygenase-like genes that regulate shoot branching in Arabidopsis and pea. Genes Dev 17:1469–1474

Soundappan I, Bennett T, Morffy N, Liang Y, Stanga JP, Abbas A, Leyser O, Nelson DC (2015) SMAX1-LIKE/D53 family members enable distinct MAX2-dependent responses to strigolactones and karrikins in Arabidopsis. Plant Cell 27:3143–3159

Stanga JP, Smith SM, Briggs WR, Nelson DC (2013) SUPPRESSOR OF MORE AXILLARY GROWTH2 1 controls seed germination and seedling development in Arabidopsis. Plant Physiol 163:318–330

Stauder R, Welsch R, Camagna M, Kohlen W, Balcke GU, Tissier A, Walter MH (2018) Strigolactone levels in dicot roots are determined by an ancestral symbiosis-regulated clade of the *PHYTOENE SYNTHASE* Gene Family. Front Plant Sci 9:255

Stirnberg P, van De Sande K, Leyser HMO (2002) MAX1 and MAX2 control shoot lateral branching in Arabidopsis. Development 129:1131–1141

Stirnberg P, Furner IJ, Ottoline Leyser H (2007) MAX2 participates in an SCF complex which acts locally at the node to suppress shoot branching. Plant J 50:80–94

Sun H, Tao J, Liu S, Huang S, Chen S, Xie X, Yoneyama K, Zhang Y, Xu G (2014) Strigolactones are involved in phosphate- and nitrate-deficiency-induced root development and auxin transport in rice. J Exp Bot 65(22):6735–6746

Takeda T, Suwa Y, Suzuki M, Kitano H, Ueguchi-Tanaka M, Ashikari M, Matsuoka M, Ueguchi C (2003) The OsTB1 gene negatively regulates lateral branching in rice. Plant J 33:513–520

Tan BC, Schwartz SH, Zeevaart JA, McCarty DR (1997) Genetic control of abscisic acid biosynthesis in maize. Proc Natl Acad Sci USA 94:12235–12240

Torres-Vera R, García JM, Pozo MJ, López-Ráez JA (2014) Do strigolactones contribute to plant defence? Mol Plant Pathol 15:211–216

Tsuchiya Y, Yoshimura M, Sato Y, Kuwata K, Toh S, Holbrook-Smith D, Zhang H, McCourt P, Itami K, Kinoshita T (2015) Probing strigolactone receptors in *Striga hermonthica* with fluorescence. Science 349:864–868

Ueguchi-Tanaka M, Matsuoka M (2010) The perception of gibberellins: clues from receptor structure. Curr Opin Plant Biol 13:503–508

Ueno K, Nomura S, Muranaka S, Mizutani M, Takikawa H, Sugimoto Y (2011) Ent-2-'-epi-orobanchol and its acetate, as germination stimulants for *Striga gesnerioides* seeds isolated from cowpea and red clover. J Agric Food Chem 59:10485–10490

Ueno K, Furumoto T, Umeda S, Mizutani M, Takikawa H, Batchvarova R, Sugimoto Y (2014) Heliolactone, a non-sesquiterpene lactone germination stimulant for root parasitic weeds from sunflower. Phytochemistry 108:122–128

Umehara M, Hanada A, Yoshida S, Akiyama K, Arite T, Takeda-Kamiya N, Magome H, Kamiya Y, Shirasu K, Yoneyama K, Kyozuka J, Yamaguchi S (2008) Inhibition of shoot branching by new terpenoid plant hormones. Nature 455:195–200

Umehara M, Hanada A, Magome H, Takeda-Kamiya N, Yamaguchi S (2010) Contribution of strigolactones to the inhibition of tiller bud outgrowth under phosphate deficiency in rice. Plant Cell Physiol 51:1118–1126

Van Ha C, Leyva-González MA, Osakabe Y, Tran UT, Nishiyama R, Watanabe Y, Tanaka M, Seki M, Yamaguchi S, Van Dong N (2014) Positive regulatory role of strigolactone in plant responses to drought and salt stress. Proc Natl Acad Sci USA 111:851–856

Vishwakarma K, Upadhyay N, Kumar N, Yadav G, Singh J, Mishra RK, Kumar V, Verma R, Upadhyay R, Pandey M (2017) Abscisic acid signaling and abiotic stress tolerance in plants: a review on current knowledge and future prospects. Front Plant Sci 8:161

Wallner E-S, López-Salmerón V, Belevich I, Poschet G, Jung I, Grünwald K, Sevilem I, Jokitalo E, Hell R, Helariutta Y (2017) Strigolactone-and karrikin-independent SMXL proteins are central regulators of phloem formation. Curr Biol 27:1241–1247

Walter MH (2013) Role of carotenoid metabolism in the arbuscular mycorrhizal symbiosis. In: Molecular microbial ecology of the rhizosphere, vol 1 & 2. Wiley, Hoboken, NJ, pp 513–524

Walter MH, Strack D (2011) Carotenoids and their cleavage products: biosynthesis and functions. Nat Prod Rep 28:663–692

Wang Y, Bouwmeester HJ (2018) Structural diversity in the strigolactones. J Exp Bot 69:2219–2230

Wang H, Wang H (2015) The miR156/SPL module, a regulatory hub and versatile toolbox, gears up crops for enhanced agronomic traits. Mol Plant 8:677–688

Wang L, Wang B, Jiang L, Liu X, Li X, Lu Z, Meng X, Wang Y, Smith SM, Li J (2015) Strigolactone signaling in Arabidopsis regulates shoot development by targeting D53-like SMXL repressor proteins for ubiquitination and degradation. Plant Cell 27:3128–3142

Waters MT, Brewer PB, Bussell JD, Smith SM, Beveridge CA (2012a) The Arabidopsis ortholog of rice DWARF27 acts upstream of MAX1 in the control of plant development by strigolactones. Plant Physiol 159:1073–1085

Waters MT, Nelson DC, Scaffidi A, Flematti GR, Sun YK, Dixon KW, Smith SM (2012b) Specialisation within the DWARF14 protein family confers distinct responses to karrikins and strigolactones in Arabidopsis. Development 139:1285–1295

Waters MT, Scaffidi A, Flematti GR, Smith SM (2013) The origins and mechanisms of karrikin signalling. Curr Opin Plant Biol 16:667–673

Waters MT, Scaffidi A, Sun YK, Flematti GR, Smith SM (2014) The karrikin response system of Arabidopsis. Plant J 79:623–631

Waters MT, Gutjahr C, Bennett T, Nelson DC (2017) Strigolactone signaling and evolution. Annu Rev Plant Biol 68:291–322

Wen C, Zhao Q, Nie J, Liu G, Shen L, Cheng C, Xi L, Ma N, Zhao L (2016) Physiological controls of chrysanthemum DgD27 gene expression in regulation of shoot branching. Plant Cell Rep 35:1053–1070

Werck-Reichhart D, Feyereisen R (2000) Cytochromes P450: a success story. Genome Biol 1: Reviews3003. 3001

Xie X (2016) Structural diversity of strigolactones and their distribution in the plant kingdom. J Pestic Sci 41:175–180

Xie X, Kusumoto D, Takeuchi Y, Yoneyama K, Yamada Y, Yoneyama K (2007) 2-′-Epi-orobanchol and solanacol, two unique strigolactones, germination stimulants for root parasitic weeds, produced by tobacco. J Agri Food Chem 55:8067–8072

Xie X, Yoneyama K, Kusumoto D, Yamada Y, Takeuchi Y, Sugimoto Y, Yoneyama K (2008) Sorgomol, germination stimulant for root parasitic plants, produced by *Sorghum bicolor*. Tetrahedron Lett 49:2066–2068

Xie X, Yoneyama K, Harada Y, Fusegi N, Yamada Y, Ito S, Yokota T, Takeuchi Y, Yoneyama K (2009) Fabacyl acetate, a germination stimulant for root parasitic plants from *Pisum sativum*. Phytochemistry 70:211–215

Xie X, Yoneyama K, Yoneyama K (2010) The strigolactone story. Annu Rev Phytopathol 48:93–117

Xie X, Yoneyama K, Kisugi T, Nomura T, Akiyama K, Asami T, Yoneyama K (2015) Strigolactones are transported from roots to shoots, although not through the xylem. J Pestic Sci 40:214–216

Xie X, Kisugi T, Yoneyama K, Nomura T, Akiyama K, Uchida K, Yokota T, McErlean CS, Yoneyama K (2017) Methyl zealactonoate, a novel germination stimulant for root parasitic weeds produced by maize. J Pestic Sci 42:58–61

Yao R, Ming Z, Yan L, Li S, Wang F, Ma S, Yu C, Yang M, Chen L, Chen L, Li Y, Yan C, Miao D, Sun Z, Yan J, Sun Y, Wang L, Chu J, Fan S, He W, Deng H, Nan F, Li J, Rao Z, Lou Z, Xie D (2016) DWARF14 is a non-canonical hormone receptor for strigolactone. Nature 536:469–473

Yao R, Wang F, Ming Z, Du X, Chen L, Wang Y, Zhang W, Deng H, Xie D (2017) ShHTL7 is a non-canonical receptor for strigolactones in root parasitic weeds. Cell Res 27(6):838–841

Yasuda N, Sugimoto Y, Kato M, Inanaga S, Yoneyama K (2003) (+)-Strigol, a witchweed seed germination stimulant, from Menispermum dauricum root culture. Phytochemistry 62:1115–1119

Yokota T, Sakai H, Okuno K, Yoneyama K, Takeuchi Y (1998) Alectrol and orobanchol, germination stimulants for *Orobanche minor*, from its host red clover. Phytochemistry 49:1967–1973

Yoneyama K, Xie X, Kusumoto D, Sekimoto H, Sugimoto Y, Takeuchi Y, Yoneyama K (2007) Nitrogen deficiency as well as phosphorus deficiency in sorghum promotes the production and exudation of 5-deoxystrigol, the host recognition signal for arbuscular mycorrhizal fungi and root parasites. Planta 227:125–132

Yoneyama K, Kisugi T, Xie X, Yoneyama K (2013) Chemistry of strigolactones: why and how do plants produce so many strigolactones? In: Molecular microbial ecology of the rhizosphere, vol 1 & 2. Wiley, Hoboken, NJ, pp 373–379

Yoneyama K, Arakawa R, Ishimoto K, Kim HI, Kisugi T, Xie X, Nomura T, Kanampiu F, Yokota T, Ezawa T (2015) Difference in striga-susceptibility is reflected in strigolactone secretion profile, but not in compatibility and host preference in arbuscular mycorrhizal symbiosis in two maize cultivars. New Phytol 206:983–989

Yoneyama K, Mori N, Sato T, Yoda A, Xie X, Okamoto M, Iwanaga M, Ohnishi T, Nishiwaki H, Asami T (2018) Conversion of carlactone to carlactonoic acid is a conserved function of MAX 1 homologs in strigolactone biosynthesis. New Phytol 218:1522–1533

Zhang Y, van Dijk AD, Scaffidi A, Flematti GR, Hofmann M, Charnikhova T, Verstappen F, Hepworth J, van der Krol S, Leyser O, Smith SM, Zwanenburg B, Al-Babili S, Ruyter-Spira C, Bouwmeester HJ (2014) Rice cytochrome P450 MAX1 homologs catalyze distinct steps in strigolactone biosynthesis. Nat Chem Biol 10:1028–1033

Zhang Y, Cheng X, Wang Y, Diez-Simon C, Flokova K, Bimbo A, Bouwmeester HJ, Ruyter-Spira C (2018) The tomato MAX1 homolog, SlMAX1, is involved in the biosynthesis of tomato strigolactones from carlactone. New Phytol 219(1):297–309

Zhao LH, Zhou XE, Wu ZS, Yi W, Xu Y, Li S, Xu TH, Liu Y, Chen RZ, Kovach A, Kang Y, Hou L, He Y, Xie C, Song W, Zhong D, Xu Y, Wang Y, Li J, Zhang C, Melcher K, Xu HE (2013) Crystal structures of two phytohormone signal-transducing alpha/beta hydrolases: karrikin-signaling KAI2 and strigolactone-signaling DWARF14. Cell Res 23:436–439

Zhou F, Lin Q, Zhu L, Ren Y, Zhou K, Shabek N, Wu F, Mao H, Dong W, Gan L, Ma W, Gao H, Chen J, Yang C, Wang D, Tan J, Zhang X, Guo X, Wang J, Jiang L, Liu X, Chen W, Chu J, Yan C, Ueno K, Ito S, Asami T, Cheng Z, Wang J, Lei C, Zhai H, Wu C, Wang H, Zheng N, Wan J (2013) D14-SCF(D3)-dependent degradation of D53 regulates strigolactone signalling. Nature 504:406–410

Chapter 2
Strigolactones as Plant Hormones

Catherine Rameau, Sofie Goormachtig, Francesca Cardinale, Tom Bennett, and Pilar Cubas

Abstract In the last decade strigolactones have been recognized as a novel type of plant hormones. They are involved in the control of key developmental processes such as lateral shoot outgrowth and leaf and root development, among others. In addition, strigolactones modulate plant responses to abiotic stresses like phosphate starvation and drought. Here we summarize the current knowledge of the widely conserved functions of strigolactones in the control of plant development and stress responses as well as some of their reported species-specific roles. In addition, we will review their known genetic and functional interactions with other phytohormones. The newly discovered activities of strigolactones as plant hormones raise the possibility of using these compounds and their signalling pathways as tools to optimise species of agronomical importance.

Keywords Shoot branching · Root development · Response to abiotic stress · Hormone signalling cross-talk

C. Rameau
Institut Jean-Pierre Bourgin, INRA, AgroParisTech, CNRS, Université Paris-Saclay, Versailles, France
e-mail: Catherine.Rameau@inra.fr

S. Goormachtig
Department of Plant Biotechnology and Bioinformatics, Ghent University, Ghent, Belgium

Center for Plant Systems Biology, VIB, Ghent, Belgium
e-mail: sogoo@psb.vib-ugent.be

F. Cardinale
Plant Stress Lab, DISAFA—University of Turin, Grugliasco (TO), Italy
e-mail: francesca.cardinale@unito.it

T. Bennett
School of Biology, University of Leeds, Leeds, UK
e-mail: t.a.bennett@leeds.ac.uk

P. Cubas (✉)
Plant Molecular Genetics Department, Centro Nacional de Biotecnología (Consejo Superior de Investigaciones Científicas), Campus Universidad Autónoma de Madrid, Madrid, Spain
e-mail: pcubas@cnb.csic.es

H. Koltai, C. Prandi (eds.), *Strigolactones - Biology and Applications*,
https://doi.org/10.1007/978-3-030-12153-2_2

2.1   Introduction

Plants adjust their growth and development to changing environmental and endogenous conditions. This plasticity is partly achieved by their capability to perceive external and internal cues in various plant organs and to convey this information over long distances to coordinate adaptive responses throughout the plant. This communication is primarily mediated by a small group of chemical messengers, the 'phytohormones', which are transported systemically within the plant body. Inside cells, phytohormones act at low concentrations; after binding to their receptors, they activate signalling cascades, trigger changes in gene expression and ultimately activate developmental and metabolic programs that allow plants to adapt to local environmental conditions.

Until 2008, eight classes of phytohormones were recognized: auxin, cytokinin (CK), abscisic acid (ABA), gibberellic acid (GA), ethylene, brassinosteroids, jasmonic acid and salicylic acid. Strigolactones (SL) had been known for the previous 50 years as root-exuded compounds that stimulated the germination of parasitic plant seeds (Cook et al. 1966). Much later, SLs were reported to promote branching of arbuscular mycorrhizal (AM) fungi, providing a convincing explanation for their exudation by plants (Akiyama et al. 2005; Parniske 2008). Finally, it became clear in 2008 that SLs also acted as long-distance endogenous signals within the plant (Gomez-Roldan et al. 2008; Umehara et al. 2008), controlling plant architecture and mediating adaptive responses to stress. SLs thus became recognized as a novel class of phytohormones.

The molecular mechanisms of SL signalling resemble those of other phytohormones such as jasmonic acid, auxin and GA: binding of a SL molecule to its receptor triggers polyubiquitination and proteasomal degradation of the pathway repressors. Key components of SL perception and signalling have been characterized (reviewed in Waters et al. 2017). The SL receptor is an α–β hydrolase (D14, in rice, Arite et al. 2009), widely conserved in flowering plants (e.g. Hamiaux et al. 2012; Chevalier et al. 2014). Upon SL binding and hydrolysis, D14 undergoes conformational changes that facilitate its interaction with an F-box protein, MAX2 in *Arabidopsis* (Stirnberg et al. 2007), component of an SKP1-CUL1-F-box-protein (SCF)-type ubiquitin ligase complex. Upon interaction with D14, MAX2 promotes recognition, ubiquitination and degradation of the SL pathway repressors: D53 in rice (Jiang et al. 2013; Zhou et al. 2013), the D53 orthologs SMXL6, SMXL7 and SMXL8 in *Arabidopsis* (Soundappan et al. 2015; Wang et al. 2015). D53-like repressors contain domains that allow them to interact with transcriptional corepressors like TOPLESS, which in turn interact with the histone deacetylase complex and promote chromatin remodelling and gene silencing. In summary, SL signalling leads to degradation of D53-like corepressors, which elicits SL-related transcriptional responses. In parallel, in response to SL the D14 receptor is also ubiquitinated and degraded in *Arabidopsis* and rice, although at a slower rate than D53 proteins (Chevalier et al. 2014; Hu et al. 2017).

SLs have now been implicated in controlling a wide range of morphological traits, including shoot branching; leaf shape; leaf senescence; internode growth; shoot branching angle and stem secondary growth (Sect. 2.2); growth of primary, lateral and adventitious roots; and growth of root hairs (Sect. 2.3). SLs are also involved in the adaptation of plants to abiotic stresses such as phosphate starvation and drought (Sect. 2.4). Although some of these roles have been confirmed in a large number of species (e.g. thale cress, garden pea, petunia, rice) which indicates that their functions are widely conserved across angiosperms, other functions seem to be specific to just a few species analysed so far.

SLs do not regulate these processes alone, but in concert with other hormones, either antagonistically or synergistically (Sect. 2.5). In addition, SLs can cross-regulate the activity of some phytohormones, by controlling their biosynthesis, stability and signal transduction. To understand how SL signalling interacts with the activity of other plant hormones is a current challenge essential to elucidate their role in the control of plant growth and stress responses.

2.2   SLs and the Regulation of Shoot Architecture

2.2.1   SLs Repress Shoot Branching

Shoot branching results from the outgrowth of axillary buds formed in the axils of many leaves (Fig. 2.1). During postembryonic development, axillary meristems are initiated, and many will go on to develop a few leaves before arresting to form an axillary bud. These buds may either remain dormant or grow to generate a branch, which is a secondary shoot axis with similar features to the main stem. Different patterns of shoot branching can be observed depending on environmental, genetic and developmental factors. For example, at floral transition, axillary buds that were

Fig. 2.1 SLs suppress branch outgrowth. Node of a pea SL-deficient mutant *ramosus1*, in which an axillary bud has grown out into a branch. In wild-type pea plants (not shown), axillary buds remain dormant

dormant may start to grow out. Each axillary bud integrates many endogenous and external signals to take a decision on whether to activate or not. SLs are one of the key signals that repress the outgrowth of axillary buds.

Shoot branching was intensively studied during the 'golden era' of plant physiology in the 1920s and 1930s, leading to the development of the classical theory of apical dominance (Box 2.1). Auxin was known to be a key signal regulating apical dominance, but its mechanism of action remained unclear for many years. In the 1990s and early 2000s, a genetic approach to understanding shoot branching developed in several labs led to the isolation of high shoot branching mutants in pea (*Pisum sativum*, *ramosus* (*rms*) mutants), petunia (*Petunia x hybrida*, *decreased apical dominance* (*dad*) mutants), thale cress (*Arabidopsis thaliana* (*Arabidopsis*), *more axillary growth* (*max*) mutants (Figs. 2.1, 2.2)) and rice (*Oryza sativa, dwarf* (*d*) or *high-tillering dwarf* (*htd*) mutants) (Simons et al. 2006; Beveridge et al. 2009; Wang and Li 2011; Waldie et al. 2014). Most of the genes identified through these mutants were subsequently shown to be components in the SL biosynthesis or signalling pathways (see below).

> **Box 2.1 The Classical Theory of Apical Dominance**
> Apical dominance is the term used to describe the control of the shoot tip over axillary bud outgrowth. Removal of the shoot tip (decapitation) is a technique well known by gardeners to make a plant bushier as it stimulates axillary bud outgrowth. This technique was also commonly used to study axillary bud outgrowth. The shoot tip being the main source of auxin, it was postulated that auxin synthesized in the shoot apex and transported rootwards acted indirectly by regulating the transport or the levels of other signals ('second messengers') moving shootwards. Auxin negatively regulates cytokinin levels and positively SL levels. Thus cytokinin and SL could be considered as second messengers by which auxin represses axillary bud outgrowth. Auxin might also act in the main stem to repress auxin export from the axillary bud and consequently, its outgrowth. More recently, the involvement of sugars acting as signalling molecules in apical dominance was proposed. Using plants with long internodes, it was shown that after removal of the shoot apex, the decrease in auxin content along the stem was too slow to explain the early growth of axillary buds located at basal nodes. The strong demand for sugar by the shoot tip may participate in the apical dominance. After decapitation, sugars are redistributed very rapidly to trigger bud outgrowth (Mason et al. 2014). The relationship between sugars and SLs is still unclear.

In pea, these high shoot branching mutants were first characterized using simple grafting experiments. These experiments led to the conclusion that some mutants were deficient in a novel branching inhibitor and others were not responding to the branching inhibitor (Box 2.2). A major clue for the discovery of this unknown branching inhibitor came from the cloning of the *MAX3* and *MAX4* genes from *Arabidopsis*, which both encoded CAROTENOID CLEAVAGE DIOXYGENASE enzymes (CCD7 and CCD8, respectively) (Sorefan et al. 2003; Booker et al. 2004)

Fig. 2.2 Phenotypes of SL-synthesis mutants. (**a**) Shoot branching phenotype of *Arabidopsis* wild-type plant (left) and SL-synthesis mutant *max4* (right). (**b**) The wild-type pea plant (left) has fewer branches with larger branch angles than those of the SL-deficient *rms1* mutant, which displays more vertical branches. (**c**) Petunia SL-deficient mutant *dad1-1* (right) is bushier and later flowering than wild type (left) (Photo kindly provided by Kim Snowden)

(Fig. 2.2a). These results suggested that the branching inhibitor was very likely a carotenoid-derived compound. At the same time, a group working on parasitic plants showed that SLs, molecules already known to stimulate seed germination of these plants, were carotenoid derived (Matusova et al. 2005). Demonstration that SLs were also the elusive branching inhibitor involved exogenous applications of *racemic*-GR24 (*rac*-GR24), a synthetic SL (Box 2.3), to test whether it inhibited shoot branching. Indeed, *rac*-GR24 applications inhibited axillary bud outgrowth in SL synthesis mutants but not in response mutants, establishing that SLs were indeed the mystery branching regulators (Gomez-Roldan et al. 2008; Umehara et al. 2008). It is worth mentioning that *rac*-GR24 consists of a mixture of enantiomers, one of which is a structural analogue of natural SLs, but another activates a different signalling pathway (Box 2.4) (Scaffidi et al. 2014).

Box 2.2 Characterization of High Branching Mutants Using Grafting: Evidence of a Novel Branching Inhibitor

A branching inhibitor can move from the wild-type root to the shoot to inhibit the branching phenotype of the synthesis mutant (Graft 1), but not of the response mutant (Graft 2). The branching inhibitor also stimulates the height of the plant (Graft 1). The branching inhibitor is synthesized in the wild-type root and also in the wild-type shoot (Graft 3). The branching inhibitor can only move in a root-to-shoot direction (Graft 4) as the branching inhibitor from the wild-type shoot cannot inhibit the branching of the synthesis-mutant shoot. Colours indicate the genotype of the plant.

Box 2.3 *Rac*-GR24, a Widely Used Synthetic SL Analogue

Legend: Left-hand structures: The racemic solution of GR24, the most commonly used synthetic analogue of SL, is composed by the equimolar mixture of the two enantiomers GR24^{5DS} and GR24$^{ent\text{-}5DS}$. Right-hand: Molecular structures of strigol and orobanchol, two naturally occurring SLs characterized by *beta* and *alpha* orientation of the C ring, respectively. They are representative of the two main molecular types of natural strigolactones; both share the R configuration at the C-2'. GR24^{5DS} is a structural analogue of strigol, but GR24$^{ent\text{-}5DS}$ is not an analogue of either strigol- or orobanchol-type SLs.

Box text: The influence of a particular hormone on plant development is often studied by using synthetic analogues, since naturally occurring

(continued)

Box 2.3 (continued)

phytohormones are often difficult to obtain at high quantities. For SL research, the most commonly used analogue is *racemic*-GR24 (*rac*-GR24). GR refers to the initials of the chemist Gerald Roseberry, who originally synthesized the molecule. It is important to keep in mind that *rac*-GR24 consists of a mixture of enantiomers, of which only GR24^{5DS}, mimics the stereochemistry of the natural SLs, while GR24$^{ent-5DS}$ activated a separate signalling pathway, the KAI2 pathway (Box 2.4) (Scaffidi et al. 2014). In several early studies, the use of *rac*-GR24 led to erroneously propose a role for SLs in processes probably controlled by the yet unknown 'KAI2 ligand' (KL) (Box 2.4).

Box 2.4   SLs and 'KAI2 Ligand'

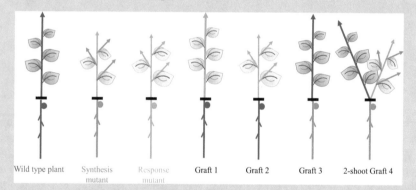

| Wild type plant | Synthesis mutant | Response mutant | Graft 1 | Graft 2 | Graft 3 | 2-shoot Graft 4 |

One of the most unexpected aspects of the SL story is that its discovery led to the fortuitous finding of a second signalling pathway, which appears to act in the perception of another novel phytohormone. The *max2* SL signalling mutant from *Arabidopsis* was known for some time to have additional and unexplained phenotypes relative to SL synthesis mutants. Eventually it became clear that these other phenotypes were related to a second signalling pathway that uses the SCF^{MAX2} ubiquitin ligase to degrade target proteins (Nelson et al. 2011). The KAI2 α/β hydrolase protein was identified as the receptor for this second pathway and is a close homologue of the D14 protein that acts as a SL receptor (Waters et al. 2012) (Chap. 1). KAI2 perceives exogenous smoke-derived molecules (karrikins) but is also inferred from a number of approaches to perceive an endogenous phytohormone currently referred to as 'KAI2 ligand' (KL) (Conn and Nelson 2016; Sun et al. 2016a). The probable proteolytic targets of KAI2 signalling (SMAX1 and SMXL2 in *Arabidopsis*) are also close homologues of the proteolytic targets of SL

(continued)

Box 2.4 (continued)

signalling (SMXL6, SMXL7 and SMXL8 in *Arabidopsis*, D53 in rice). Intriguingly, the KAI2 signalling pathway appears to be much more ancient that the SL signalling pathway, even though SL synthesis itself is ancient (Bythell-Douglas et al. 2017; Walker and Bennett 2017). SL signalling thus appears to have arisen by duplication and divergence of KL signalling components (Walker and Bennett 2017).

In a further coincidence, the most widely used synthetic SL analogue, GR24 (Box 2.3), transpired to a racemic mixture of four molecules two of which trigger D14 signalling and two of which trigger KAI2 signalling (Scaffidi et al. 2014). The use of racemic-GR24 (*rac*-GR24) thus leads to effects that may not be SL related, causing considerable problems for interpretation of data. In many early studies, the use of *rac*-GR24 led to erroneous conclusions about the roles of SLs in several processes (which may actually rather be controlled by KL). These problems were often confounded by the use of *max2* mutants, which are not specifically SL signalling mutants but also KL signalling mutants and which were again used to conclude that SL controlled various processes. Throughout this chapter, we have tried to highlight where there is doubt about the involvement of SLs in a given process due to these *rac*-GR24/*max2* issues.

Two main hypotheses, not mutually exclusive, have been developed to explain the mechanisms by which SL interacts with auxin to regulate axillary bud outgrowth (Waters et al. 2017). One mechanism proposes that SLs regulate auxin transport (Box 2.5, Sect. 2.5). The other mechanism proposes that SLs promote expression of the *BRANCHED1 (BRC1)* gene, a growth suppressor that encodes a TCP transcription factor expressed in the axillary buds of many plant species (e.g. Aguilar-Martínez et al. 2007; Martín-Trillo et al. 2011; Braun et al. 2012). *BRC1* is a homologue of the maize *Teosinte branched1 (Tb1)* gene, known for its role in maize domestication (Doebley et al. 1997). *BRC1*- and *TB1*-related genes partly mediate the SL signalling pathway in the control of axillary bud growth. For instance, *rac*-GR24 treatments do not inhibit axillary bud outgrowth in the pea *brc1* or *Ostb1* mutants (Minakuchi et al. 2010; Braun et al. 2012). However, genetic analysis in *Arabidopsis* shows that the effects of *BRC1* and SLs are also partially independent (Seale et al. 2017). In pea, *BRC1* expression was shown to be repressed by cytokinin and sucrose, both of which act antagonistically with SLs in the control of axillary bud outgrowth (Braun et al. 2012; Mason et al. 2014).

Box 2.5 Auxin Transport and Shoot Branching

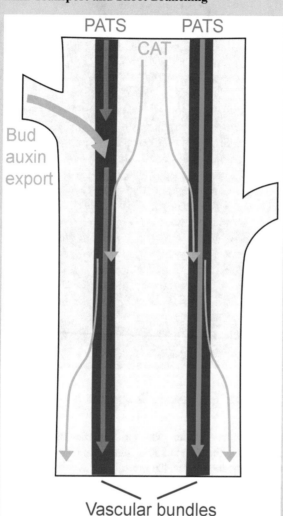

Legend: Schematic diagram illustrating the polar auxin transport stream (PATS) and connective auxin transport (CAT) in a plant stem. The PATS (pink arrows) occurs through the vascular cambium and xylem parenchyma cells within the vascular bundles (dark blue) (but not in xylem or phloem vessels). Auxin from other tissues in the stem connects to the PATS through CAT (green arrows), and auxin in the PATS can be dispersed to other tissues through CAT. Buds that are beginning to grow and export auxin also connect to the PATS through CAT.

(continued)

Box 2.5 (continued)

Text: Plants have highly specific mechanisms for the regulated transport of auxin around the plant body, a process that strongly enhances the informational nature of the auxin signal. Auxin is predominantly uncharged at apoplastic pH and can move freely into cells through the plasma membrane. However, at cellular pH, auxin is predominantly negatively charged and can only exit cells by being actively exported by efflux proteins of the ABCB or PIN families. Thus, the direction of auxin transport can be controlled at the level of cellular efflux and can range from non-polar to highly polarized, depending on the subcellular localization of efflux proteins. Transport can range in scale from localized flux within tissues to broader flux between tissues. Systemic bulk auxin transport, connecting all parts of the plant, also occurs in a predominantly shoot-to-root direction. This high-conductance 'polar auxin transport stream' (PATS) is associated with vascular tissues but does not occur in the vasculature itself. Tissues are connected to the PATS through 'connective auxin transport' (CAT), which is less polar and lower conductance in nature. Regulation of CAT and PATS is important for determining information flow in the plant and, as such, is important for controlling various aspects of development. For instance, as new shoot branches activate, they begin to export large amounts of auxin, which is correlated with their ability to grow. Indeed, there is substantial evidence that sustained auxin transport is necessary for branch outgrowth. Thus, shoot branching can ultimately be controlled by altering the strength and polarity of both CAT and PATS.

2.2.2 SLs Promote Internode Elongation and Height

If a high shoot branching is the most striking phenotype of SL-deficient and signalling mutants, another remarkable phenotype of these mutants is their semidwarfism and short internodes. To investigate whether this phenotype resulted from the diversion of energy from the main stem to lateral shoots, axillary buds or branches of pea plants were manually removed in SL-deficient mutants. This treatment had no significant effect on internode length: SL mutants remained dwarf even when the plants had only one main stem. These results indicated that the reduced stature of SL-deficient mutants was not simply a consequence of their increased branching (de Saint Germain et al. 2013). Furthermore, in barrel medic (*Medicago truncatula*, Medicago), in which wild-type plants are already highly branched, SL-deficient mutants still have shorter internodes despite no further increase in shoot branching (Lauressergues et al. 2015).

The molecular mechanisms involved in SL action on the internode elongation are not yet clearly understood. The phytohormone GA promotes internode elongation and acts both on cell division and cell elongation. In pea, it was shown that SLs stimulate internode elongation by affecting cell division but not cell elongation and that they acted independently of GA. Consistently, when mutants with altered SL and GA levels were combined, additive effects were observed on internode elongation (de Saint Germain et al. 2013).

2.2.3 Effect of SLs on Gravitropic Set Point Angle

The angle between branches and the main stem is an important component of shoot architecture and yield. In crops, a compromise has to be found between a broad-spreading plant that occupies too much space in the field and a very compact plant, which will be less efficient in light harvesting and more susceptible to pathogen attacks. SL mutants in pea and *Arabidopsis* have strongly reduced branch angles (Fig. 2.2b) (Liang et al. 2016). Shoot organs naturally display negative gravitropism (i.e. reorientation away from gravity), and maintenance of branch angles at non-vertical angles requires 'offsetting' the gravitropic stimulus. This results in a gravitropic set point angle (GSA), which an organ will return to when gravi-stimulated (Box 2.6).

> **Box 2.6 Plant Gravitropism**
> Plants are able to sense gravity, and their shoots generally grow away from gravitropic stimuli (negative gravitropism), while their roots generally grow towards gravity (positive gravitropism). Gravity is perceived in specific cells in the shoot and root (statocytes), by the sedimentation of starch-accumulating amyloplasts (statoliths). This process leads to redistribution of the phytohormone auxin and a differential growth between the lower and upper sides of the shoot or root.

It is currently unclear where SLs fit into the regulation of GSA. Several genes are known to control GSA, such as members of the *LAZY1* gene family. The *lazy1* mutants in rice and *Arabidopsis* have defective shoot gravitropism and display a spreading phenotype with large branch angles. In *Arabidopsis*, three *LAZY* genes, acting in specific endodermal cells in the shoots (the statocytes), have a key role in the generation of the asymmetric distribution of auxin between the lower and upper sides of the stem, which is required for gravitropic response (Taniguchi et al. 2017). In rice, a genetic screen aimed at finding suppressors of *lazy1*, led to the identification of mutants affected in SL biosynthesis or SL signalling (Sang et al. 2014). However, SL regulation of shoot gravitropism was proposed to be *LAZY1*-independent. While *LAZY1* stimulates lateral auxin transport to generate an asymmetric distribution of auxin, SLs attenuate this auxin asymmetric distribution by repressing auxin levels in the lower side of the shoot (Sang et al. 2014). This repressing effect of SLs on auxin biosynthesis was also recently demonstrated in pea (Ligerot et al. 2017).

2.2.4 SLs Control Secondary Growth in Stems

Another key phenotype of SL-related mutants is their reduced stem diameter in comparison to wild type. Stem width results from secondary growth due to the activity of the cambium, a cylindrical meristematic tissue that divides radially to add thickness to the stem. The cambium comes in two major forms: vascular cambium, in which the vascular bundles of xylem and phloem vessels form (Box 2.7), and interfascicular cambium, which occurs between the vascular bundles. *Arabidopsis* SL mutants have a reduced cambium activity compared to wild type: cambium-specific genes are expressed at lower levels, and the stems display reduced lateral extension of tissue derived from the interfascicular cambium (Box 2.7). Local treatment of stems with *rac*-GR24 stimulates cell division in interfascicular regions, suggesting a direct effect of SLs on cambium activity (Agusti et al. 2011). Interestingly, experiments of plant decapitation not only induced axillary bud outgrowth but also reduced stem secondary growth, suggesting that regulation of branching and secondary growth are linked.

Box 2.7 Cambium and Vascular Tissue

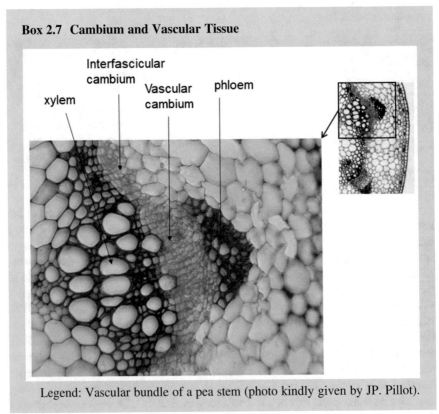

Legend: Vascular bundle of a pea stem (photo kindly given by JP. Pillot).

(continued)

> **Box 2.7** (continued)
>
> Text: Cambium is a specialized, undifferentiated layer of cells that allows the radial growth of shoot and roots. In the vascular bundles, vascular cambium forms a layer of meristematic cells located between the xylem and phloem vessels. This meristematic tissue extends between vascular bundles to form the interfascicular cambium. The resulting ring of meristematic tissue is very clearly observed in woody plants but can also be seen in herbaceous plants like *Arabidopsis*.

2.2.5 SLs Affect Leaf Shape and Leaf Serration

In *Arabidopsis*, SL-deficient and response mutants exhibit rounder rosette leaves with shorter petioles (Stirnberg et al. 2002). This phenotype is restored to wild type in plants lacking the repressors of the pathway (*smxl6,7,8* triple mutants, Soundappan et al. 2015). Furthermore, treatment with *rac*-GR24 of SL-deficient mutants grown in hydroponic culture restores leaf shape (Scaffidi et al. 2013). The compound leaves of some Medicago ecotypes such as R108 have serrated margins. In SL mutants the degree of serration is reduced. To test whether this serration phenotype was linked to SLs, *rac*-GR24 was applied to the main shoot apex of the SL-deficient mutant, and, indeed, *rac*-GR24-treated leaflets displayed deeper serrations than mock-treated leaflets (Lauressergues et al. 2015). However, in other ecotypes, e.g. Jemalong A17, *rac*-GR24 treatments did not give deeper leaf serrations, although the root phenotype responded to the treatment. This suggests that the effect of SLs on leaf serration is not a general property in Medicago.

2.2.6 SLs and Reproductive Organ Size

The role of SLs during reproductive development is still unclear although SL synthesis genes are strongly expressed in reproductive tissues. For example, *MAX4* is expressed in the *Arabidopsis* siliques (Bainbridge et al. 2005), *CCD7* is highly expressed in tomato fruits (Vogel et al. 2010) and mRNA levels of *ZmCCD8* are particularly high in the ear (female inflorescence) of maize plants (Guan et al. 2012). Flowers, fruits and seeds of SL-deficient mutants are often smaller than those of the wild type. For instance, tomato antisense *CCD8* lines showed reduced size of flowers and fruits (Kohlen et al. 2012). However, it is yet unclear whether this reduced size is due to the high branching phenotypes, which in turn result in a higher number of flowers per plant, or a direct effect of lack of SL signalling in the fruits.

2.2.7   Other Roles of SLs: Control of Flowering Time and Tuberization

In many different species, mutants and transgenic lines with reduced mRNA levels (RNA interference lines, RNAi) of SL-biosynthesis genes (usually *CCD7* or *CCD8* orthologs) have been studied. Some of them display species-specific phenotypes additional to the general ones described above. In petunia, for instance, reduced SL synthesis leads to late flowering phenotypes. The *dad1-1* mutant, which has a transposon insertion in *PhCCD8*, has a delayed flowering time relative to wild type, as do RNAi lines of *DAD1* (Snowden et al. 2005) (Fig. 2.2c). This strong phenotype is not observed in the corresponding mutants of *Arabidopsis*, pea or rice; only a short delay in flowering time is detectable in some of these mutants. A SL-defective tomato and bird's-foot trefoil (*Lotus japonicus*) plants display severe defects in flower and fruit setting (Kohlen et al. 2012; Liu et al. 2013; Vogel et al. 2010).

Potato *StCCD8* RNAi lines provided new insights in the diversity of SL functions. These RNAi lines have altered phenotypes similar to those observed in other species (i.e. more branching, reduced height), but, in addition, they displayed additional phenotypes in tuber formation and development. Potato tubers originate from underground shoots called stolons, by swelling of the stolon tips. Stolons usually grow diageotropically, that is, perpendicular to gravity. Interestingly, stolons of the RNAi lines with reduced function of *StCCD8* tend to emerge from the soil and form aerial shoots (Pasare et al. 2013). Further investigation is needed to test whether this effect is related to the mechanism by which SLs control branching angle (see above). In addition these RNAi lines do not form flowers. Instead, basal nodes of these plants develop miniature tubers that also produce shoots. Underground, the tubers of these lines have an elongated and knobby shape and new tubers are formed directly from the mother tuber. Tubers are higher in number but smaller in size, so that the total tuber yield of the *StCCD8* RNAi plants is up to threefold lower than that of control plants. During storage, when tubers are normally dormant, *StCCD8* RNAi tubers have a higher degree of sprouting than controls (Pasare et al. 2013).

2.3   SLs and the Regulation of Root Architecture

2.3.1   SLs and Root Length

SLs have been proposed to regulate the growth of the primary root (Ruyter-Spira et al. 2011). Various SL mutants from different species, such as *Arabidopsis*, rice and barley (*Hordeum vulgare*) have a slightly shortened main root. Studies in *Arabidopsis* have suggested that this decreased length was due to a smaller root meristem than in wild-type plants. Consistent with these data, treatments with *rac*-GR24 increase the main root length in wild-type plants of several plant species. However, it is worthwhile noting that, for instance, in rice, this effect is influenced by the concentrations of phosphate or nitrate in the medium (Sun et al. 2014, see also

Sect. 2.4). Furthermore, this effect is not general for all plant species. In Medicago and tomato, *rac*-GR24 has no impact on the main root length. Moreover, in *Lotus japonicus*, transgenic lines in which the SL biosynthesis gene *CCD7* was silenced had longer primary roots than those of the wild type (Koltai et al. 2010; Liu et al. 2013; De Cuyper et al. 2015). It is currently unclear why there are differences between species. The impact of SLs on the main root length is subtle, in contrast to the much more pronounced effect of other plant hormones, such as auxins, through which SLs probably act (Matthys et al. 2016). Hence, because several other hormones control the length of the main root, the effect of SLs might become masked due to differences in the general hormone landscape between different plant species or when grown under different nutrient conditions.

2.3.2 Strigolactones in Lateral Root Development

SLs have also been proposed to exert an effect on the development of lateral roots in many species (e.g. Ruyter-Spira et al. 2011; Kapulnik et al. 2011a; Sun et al. 2014; De Cuyper et al. 2015). *Arabidopsis max2* mutants, with impaired SL (but also KAI2-related) signalling, have a phenotype of enhanced lateral root density (i.e. number of lateral roots divided by total root length). This phenotype is reduced in *smxl6,7,8* triple mutants lacking the SL signalling repressors (Soundappan et al. 2015), which supports a role of SL in this process. However, whereas in rice SL biosynthesis mutants consistently have an enhanced lateral root density phenotype, *Arabidopsis max3* and *max4* biosynthesis mutants do not display clear defects on lateral root development (Sun et al. 2014; Kapulnik et al. 2011a, b; Ruyter-Spira et al. 2011).

In addition, in *Arabidopsis* and rice, treatments with *rac*-GR24 cause reduced lateral root density in wild-type plants (Fig. 2.3). In *Arabidopsis* this response has been shown to be mainly due to inhibition of the lateral root outgrowth in the upper part of the root (Jiang et al. 2016). This phenotype could be due to signalling through either the SL or the KL pathway (Box 2.4). Indeed not only the GR24^{5DS} enantiomer – which mimics the endogenous SL – but also the GR24$^{ent\text{-}DS}$, which mimics karrikins, leads to reduced lateral root density (Box 2.4), suggesting a role for both SL and KL in the control of lateral root development (Li et al. 2016).

2.3.3 SLs in Adventitious Root Development

SLs have also been implicated in the development of adventitious roots that originate from non-root tissues in *Arabidopsis* and pea (Rasmussen et al. 2012) and also in rice crown roots (which are developmentally equivalent). Interestingly, the effect observed was opposite in *Arabidopsis* and rice: whereas in *Arabidopsis* SL biosynthesis and signalling mutants have an enhanced capacity for adventitious rooting and the number of adventitious roots is decreased by *rac*-GR24 treatments, in rice, SL-biosynthesis mutants had shorter and fewer crown roots than the wild type (Sun et al. 2015).

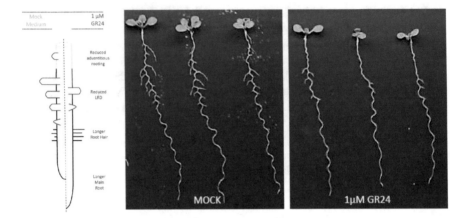

Fig. 2.3 Effects of *rac*-GR24 on the roots of *Arabidopsis* plants. Left panel: Schematic representation of the effects of *rac*-GR24 on the *Arabidopsis* root. Adapted from Matthys et al. (2016). Middle and right panels: example of *Arabidopsis* plants grown on agar plates for 8 days without (middle panel) or with (right panel) 1 µM of *rac*-GR24. Size bar is 1 cm. The effect of *rac*-GR24 on lateral root growth is illustrated. LRD, lateral root development

2.3.4 SLs in Root Hair Development

A fourth root developmental process in which SLs might be involved is root hair density and elongation. Under nutrient-limiting conditions, root hair density was lower in SL signalling and biosynthesis mutants (*max2* and *max4*, respectively) than in the wild type (Mayzlish-Gati et al. 2012) (see Sect. 2.4). Furthermore, in phosphate-deprived conditions, both SL signalling and biosynthesis mutants (*max2* and *max1*, respectively) displayed shorter root hairs than wild-type plants (Ito et al. 2015).

Consistently, *rac*-GR24 treatments cause root hair elongation in wild-type tomato and *Arabidopsis*, although not in Medicago (Koltai et al. 2010; Kapulnik et al. 2011a, b; De Cuyper et al. 2015). Detailed analyses using different enantiomers of synthetic SLs showed that compounds resembling SLs had a stronger effect on root hair growth than those expected to act through the KAI2 pathway (Box 2.4). Nevertheless, the latter also had an effect on root hair development (Artuso et al. 2015; Li et al. 2016). All these pieces of evidence support a role of SLs but probably also of the KAI2-dependent signalling pathway in root hair development.

2.3.5 Mechanism of SL Action in Roots

What do we presently know about the molecular effects caused by SLs in roots? Since SLs-auxin interactions have been demonstrated in the control of shoot branching (see Sect. 2.5), these interactions have also been studied in roots. It has been found that each hormone influences the expression of genes involved in the biosynthesis or signalling of the other. For instance, exogenous auxin treatments lead to increased transcript levels of the SL biosynthesis genes *MAX3* and *MAX4* in *Arabidopsis*. Conversely, treatments with *rac*-GR24 decrease the expression of the auxin-responsive gene *INDOLE-3-ACETIC ACID INDUCIBLE 1* (*IAA1*) (Foo et al. 2005; Hayward et al. 2009; Mashiguchi et al. 2009). The influence of SLs on PIN auxin transporters (Box 2.5) in the root has also been investigated. In general, as with the physiological effects of SLs, the impact of SL on PIN expression is not as clear as that observed during shoot branching (see also Sect. 2.5). In root tips, prolonged treatment with *rac*-GR24 resulted in a downregulation of *PIN1*, *PIN3*, and *PIN7* expression (Ruyter-Spira et al. 2011). In root epidermal cells (of the elongation zone), *rac*-GR24 treatments resulted in increased *PIN2* expression and enhanced PIN2 polarization in the plasma membrane (Pandya-Kumar et al. 2014). Further investigations revealed that, in contrast to the findings in the shoot, short *rac*-GR24 treatments did not induce PIN1 endocytosis from the plasma membrane, nor did they cause a reduction in the total PIN1 levels in roots cells (Shinohara et al. 2013).

In common with shoot branching, endogenous auxin levels seem to influence the impact of SLs on root development. Indeed, the auxin status of the plant greatly affects the effect of *rac*-GR24 on lateral root density: the negative effect of *rac*-GR24 could be modified into no effect or even a positive effect when the auxin levels were increased (Ruyter-Spira et al. 2011). The underlying basis for these observations is still not entirely clear (Shinohara et al. 2013).

SLs may also interrelate with CKs in the control of lateral root development. The CK receptor *Arabidopsis* Histidine Kinase 3 (AHK3) and the downstream regulators, the *Arabidopsis* Response Regulator 1 (ARR1)/ARR2 proteins, have been shown to interact with *rac*-GR24 to impact on lateral root development. Mutants in this CK signalling module were insensitive to *rac*-GR24, probably because of the negative effect of this module on the auxin pathway (Jiang et al. 2016). No interaction of SLs with auxin or CK has been found to affect the adventitious root development (Rasmussen et al. 2012).

Besides the interaction with other hormones, what do we know about the molecular aspects of SL signalling in the root? A study to investigate the effects of *rac*-GR24 treatment on the root proteome revealed that strongly upregulated proteins were involved in the production of various specialized metabolites, among them flavonols (Walton et al. 2016). However, the consequences of this upregulation are currently unclear.

In summary, SLs influence the development of root architecture at different stages, but the particular molecular mechanisms underlying these effects still need to be elucidated.

2.4 SLs and Plant Response to Abiotic Stress

Plants are sessile organisms that cannot avoid a changing and often challenging environment. The most frequent abiotic constraints (Box 2.8) that impact on plant growth and yield are limitations in mineral nutrients and water. The most limiting mineral nutrients are phosphorus (P) and nitrogen (N), normally absorbed from soil in the phosphate and nitrate forms, respectively. Water deprivation and high soil salinity lead to osmotic stress and reduced soil water potential (Box 2.9), which makes it harder for root tissues to extract water from the ground. This in turn, may lower the relative water content of plant tissues, thus affecting growth and development, and in extreme cases causing death. Plants have evolved a number of morphological and molecular strategies to escape, avoid or tolerate stress (Box 2.10). Phytohormones are central components of these responses.

Box 2.8 Abiotic Constraint

In biology, abiotic components are chemical and physical environmental factors that affect living organisms and their biological functions. All nonliving components of an ecosystem required for living organisms and that become limiting for growth—such as suboptimal atmospheric conditions, water or mineral nutrient resources—can be considered abiotic constraints.

Box 2.9 Water Potential

Water potential, usually represented by the Greek letter ψ, is the potential energy of water per unit volume relative to pure water in reference conditions. It quantifies the tendency of water to move from one area to another in a given system. This parameter has proven useful in understanding and computing water movements within soil, plants and atmosphere, as water will always move from areas of high ψ to areas of low (usually more negative) ψ.

Box 2.10 Stress Escape, Avoidance and Tolerance

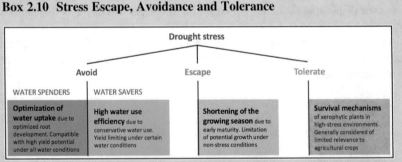

Legend: Main coping strategies of plants under stress, the example of drought. Adapted from Bodner et al. (2015) with the permission of the Editors in chief of Agronomy for Sustainable Development.

(continued)

Box 2.10 (continued)

Text: "Stress escape" is the ability of plants to complete their life cycle before the onset of stress. For example, plants do not experience drought stress if they are able to modulate their vegetative and reproductive growth according to water availability, through rapid developmental progression and plasticity (e.g. setting seeds before water is scarce). "Stress avoidance" is the ability of plants to reduce strain (i.e. the negative effects of stress) by maintaining a relatively normal metabolism and physiology under adverse environmental conditions. For instance, water content is kept at acceptable levels in plant tissues despite reduced water content in the soil, through a variety of adaptive mechanisms that allow for minimization of water loss or optimization of water uptake. "Stress tolerance" is the ability of plants to endure strain through adaptive traits. In the case of drought tolerance, these traits include maintenance of cell turgor through osmotic adjustment and other strategies (Basu et al. 2016).

Did you know that in most natural soils, bioavailable mineral nutrients are not sufficient to sustain the growth of high-yielding crop varieties and high-intensity agricultural management typical of industrialized countries. Thus, the green revolution had to rely on chemical fertilizers containing mostly N and P (Andreo-Jimenez et al. 2015), whose demand is projected to increase by an annual 3% worldwide. However, fertilizer production is energetically costly, while resources are limited and unevenly distributed. In the case of P, for instance, production rates are expected to reach their limit in a few decades due to depletion of rock phosphate stocks. In addition, the cost of N and P fertilizers has more than tripled in the last 15 years. Only 30–50% of the applied N fertilizer and around 45% of P fertilizer is taken up by crops, and a significant amount is lost from agricultural fields. In European agriculture, for example, N and P fertilizer applications average out at 150 and 90 kg/ha, respectively. However, soil leaching can affect 60% of this, leading to microbial and algal growth in water reservoirs and a negative impact on water quality and biodiversity (Tilman et al. 2002). It is generally considered that N limitation has greater impact on biomass formation, directly followed by P. However, with increasing soil age P limitation tends to become the major limiting factor of plant productivity (Czarnecki et al. 2013; Lambers et al. 2008).

Did you know that more than 800 million hectares of arable land globally is adversely affected by salinity, which is equivalent to approximately 20% of the world's cultivated land area and 50% of all irrigated lands (Sairam and Tyagi 2004). Drought is possibly the major constraint on plant growth and productivity in rain fed areas, brought about by infrequent rain and insufficient irrigation (Chaves et al. 2003).

SLs were first proposed to participate in the orchestration of stress responses based on the observation that their synthesis and exudation into soil is induced by abiotic cues. Later, a causative link between SL action and abiotic stress responses was established by the study of plants defective in SL synthesis, exudation or perception. Other experiments that used exogenous treatments with *rac*-GR24 seemed to confirm this hypothesis, but again, these must be interpreted with caution (see Box 2.4).

2.4.1   SLs and Response to Nutrient Deprivation

Nutrient availability has a strong influence on SL metabolism and distribution. In almost all plant species analysed, low P availability in the soil strongly induces SL synthesis and exudation. In some species, SL accumulation has also been observed in low N soils (Andreo-Jimenez et al. 2015; Lopez-Raez et al. 2017), but this response is thought to be due to the fact that N deprivation can induce fluctuations in the shoot P content, which in turn triggers SL accumulation and exudation at the root level (Yoneyama et al. 2012).

As SLs are known to promote beneficial symbioses (i.e. arbuscular mycorrhizal fungi and nodulating bacteria) (Lanfranco et al. 2018; Lopez-Raez et al. 2017), nutrient-responsive SL synthesis was initially thought to facilitate root colonization by symbiotic microorganisms and thus boost nutrient capture. Accordingly, it was proposed that SL synthesis in non-nodulating plants (i.e. nonlegumes) would respond to low P but not to low N. However, analysis of a wide range of plant species showed that SL induction by low N did not correlate with the plant's ability to host N-fixing bacteria; and indeed, even non-mycorrhizal plants (e.g. *Arabidopsis*) showed increased SL synthesis and exudation under low P conditions (Czarnecki et al. 2013; Mostofa et al. 2018). This implies that attracting symbiotic partners from the rhizosphere cannot be the only reason for SL accumulation under nutrient scarcity.

Plants tightly control the symbioses formed with AM fungi and rhizobia in response to external nutrient availability: under high P there is a strong suppression of mycorrhization and, under high N, suppression of nodulation. However, it must be noted that SLs are not essential for such nutrient-induced regulation of symbiosis (Lanfranco et al. 2018). For example, suppression of mycorrhization by high P cannot be overturned by *rac*-GR24 treatment (Balzergue et al. 2011; Breuillin et al. 2010). Thus, while SLs participate in the regulation of beneficial root symbioses, they are certainly not alone in this task (as discussed in Chap. 4). The extent to which SL metabolism is in turn regulated by root symbioses is also debated, because, depending on the timing after infection, extent of colonization and concurring stresses, the results differ (Lanfranco et al. 2018; Lopez-Raez et al. 2017).

What are then the biological functions of SL induction by nutritional stress? These hormones seem to mediate morphological, molecular and biochemical responses that, on one hand improve mineral uptake to avoid stress, and on the other, help the plant acclimate to nutrient deprivation independently of symbioses. For instance, SLs control shoot architectural changes in response to low nutrients. In rice and *Arabidopsis* plants grown under nutritional stress, the root/shoot biomass ratio generally increases, so that more resources are allocated to soil exploration. This reallocation is partly achieved by a decrease of shoot branching, which requires intact SL signalling (de Jong et al. 2014; Ito et al. 2015, 2016; Kohlen et al. 2011; Sun et al. 2014; Umehara et al. 2010). In addition, low P and low N promote leaf senescence and nutrient reallocation, processes also induced by SLs (Sect. 2.2) (Ueda and Kusaba 2015; Yamada et al. 2014).

Under low-nutrient conditions, root morphology is altered in a species- and nutrient-dependent way. In rice, for instance, reduced P or N lead to increased main root length and decreased lateral root density. This response is similar to that observed in plants treated with *rac*-GR24 and is compromised in SL-related mutants (Sect. 2.3) (Sun et al. 2014). In *Arabidopsis*, N deficiency leads to similar responses. In contrast, P deficiency induces a reduction of the primary root growth and an enhancement of lateral root density (Linkohr et al. 2002) that favours topsoil foraging to explore more nutrient-rich soil layers (Lynch and Brown 2001). *Arabidopsis* SL mutants grown under low P do not display increased lateral root density although they do have shorter primary roots (see also Sect. 2.3).

Other typical responses to P starvation are root hair elongation, anthocyanin accumulation, activation of phosphate transporters and acidic phosphatases (needed for efficient mobilization of soil P) and reduced plant weight. In *Arabidopsis*, *rac*-GR24 treatments induce these same responses (Sect. 2.3); and more specifically, GR24^{5DS} leads to increased root hair density under low P (Madmon et al. 2016). Conversely, plants defective in SL perception or synthesis may display delayed increase in root hair density (see above), do not accumulate anthocyanins and show reduced induction of many P-responsive genes as compared to wild type, including phosphate transporters and acidic phosphatases. These observations indicate that SLs are required for a full response to low P (de Jong et al. 2014; Ito et al. 2015; Mayzlish-Gati et al. 2012; Sun et al. 2014, 2016b, c). In rice, it was also shown that SLs are needed for proper N distribution and reallocation to different plant tissues under low-nutrient stress (Luo et al. 2018). This, together with the aforementioned role of SLs in nutrient capture, led to the proposal that SLs have a dual role in plant nutrition: on one hand, a role in direct and indirect nutrient acquisition and, on the other, a role in optimization of resource allocation (Fig. 2.4).

2.4.2 SLs and Responses to Osmotic Stress

Most of the available information on the role of SLs in responses to osmotic stress has been obtained from SL synthesis or signalling mutant plants exposed to drought, osmotic stress and/or mild salinity and from the study of the effects of treatment with exogenous SLs (*rac*-GR24) on stomata functioning. Plants with altered SL synthesis or perception are hypersensitive to both drought and osmotic stress triggered by high ionic concentration in the soil (e.g. salinity). This is due to their higher stomatal conductance (i.e. loss of gaseous water from stomata on their leaf surface) both in the absence and presence of stress, as observed in *Arabidopsis*, tomato and *Lotus*

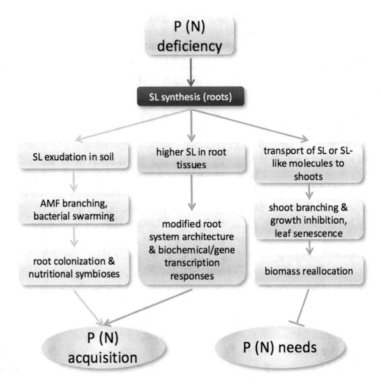

Fig. 2.4 A dual function for SLs in nutrient acquisition and allocation. Under nutrient deficiency, plants need to minimize the production of new branches to direct the limited resources to existing shoots and to maximize nutrient acquisition from soil. P (and in some species, N) deficiency stimulates SL synthesis in roots and exudation to soil. Elevated SLs (acting as endogenous hormones) act locally by modifying the architecture of the root system to increase root coverage (to explore larger soil volumes) and provide more surface (to allow for higher uptake rate). SLs are also transported shootwards to suppress shoot branching and accelerate senescence, which optimizes resource utilization. SL exudation to soil also serves as a rhizosphere signal for potential symbiotic partners, such as AM fungi and, where applicable, nodulating bacteria. This is an indirect strategy to increase nutrient acquisition. Adapted with permission from Czarnecki et al. (2013)

japonicus mutants. This phenotype is due, at least in part, to lower sensitivity to ABA and/or impaired transport of ABA. This conclusion is based on the observation that stomatal closure in response to treatments with exogenous ABA is much slower and incomplete in SL mutants, relative to their respective wild types (Ha et al. 2014; Liu et al. 2015; Visentin et al. 2016; Lv et al. 2018). However, the molecular mechanisms underpinning this are currently unknown.

Endogenous ABA content is lower in SL-deficient tomato plants. Under drought, ABA levels are even more severely reduced, which may contribute to the drought-hypersensitive phenotype of these plants (Torres-Vera et al. 2013; Visentin et al. 2016). However, in other species, SL mutants do not have reduced ABA levels: for example, SL-insensitive and SL-deficient genotypes of *Arabidopsis* and *Lotus*, respectively, contain as much ABA as the wild type (Bu et al. 2014; Liu et al. 2015).

Consistent with a positive role for SLs in water stress responses, when *Arabidopsis*, tomato or fava bean plants are treated with *rac*-GR24, their stomata will close, leading to lower gas conductance and water loss from the leaf surface (Ha et al. 2014; Visentin et al. 2016; Zhang et al. 2018). Since stomata closure in response to this mixture is disrupted both in *d14* and *kai2* mutants of *Arabidopsis*, it is clear that not only SLs (perceived by D14) but also the endogenous KL (perceived by KAI2) promotes acclimatization of *Arabidopsis* plants exposed to drought (Li et al. 2017). This explains why the drought-sensitive phenotype of *max2* is more severe than SL synthesis mutants (Box 2.4). Indeed, in early experiments, SL signalling mutants were classified as non-hypersensitive to drought compared to *max2* (Bu et al. 2014). Furthermore, stomata closure in response to *rac*-GR24 is independent of ABA signalling: ABA-related mutants respond like wild-type plants (Lv et al. 2018).

The only genes and events proven indispensable for *rac*-GR24-triggered stomata closure are, so far, *MAX2, D14, Slow Anion Channel-Associated1* (*SLAC1*) and an ABA-independent H_2O_2/NO burst in guard cells (Lv et al. 2018). In summary, while the effects of *rac*-GR24 on stomatal closure are largely ABA independent, the effect of exogenous ABA on the same feature is at least partially dependent on SLs and KL signalling (Cardinale et al. 2018).

If endogenous SLs are needed for full stomatal closure, their levels in leaves might be expected to rise in response to stress. In agreement with this, leaves treated with ABA, or under drought or osmotic stress display higher transcript levels of SL signalling and synthesis genes (Ha et al. 2014; Lv et al. 2018; Visentin et al. 2016). However, those levels are still much lower than in roots (for SL synthesis genes), and quantitative changes in SL levels cannot be detected (Visentin et al. 2016).

While all this is occurring in the leaves, what is happening in the roots, the main site of SL production? In tomato, *Lotus* and lettuce, osmotic stress decreases SL production and exudation from the roots (Aroca et al. 2013; Liu et al. 2015; Ruiz-Lozano et al. 2016; Visentin et al. 2016). Together with the parallel induction in shoots, this observation may mean that in these species, when the soil dries out, SL synthesis shifts partly from roots to shoots. It has been suggested that this shift may carry intrinsic signalling properties. Locally, in roots, it seems necessary for ABA accumulation. Indeed, ABA content of *Lotus* roots does not increase after osmotic stress if pretreated with *rac*-GR24, as it does in non-pretreated tissues (Liu et al. 2015). Systemically, a reduction of SLs flowing shootwards apparently signals to the shoot (directly or indirectly, i.e. through a second messenger) that the root is experiencing a problem. This was shown to trigger local expression of SL synthesis genes in the shoot, leading to stomatal closure (Visentin et al. 2016). This working model of organ-specific dynamics and action of SLs in response to drought has been proposed in tomato (Fig. 2.5) and is consistent with data from other dicots. However, it remains to be tested in other plant species; for example, the SL-ABA pathways seem to interact in a different way in rice under drought (Haider et al. 2018). In this species, drought increases root SL content and exudation. Additionally, SL synthesis mutants of rice contain more ABA than wild-type plants and are more, not less

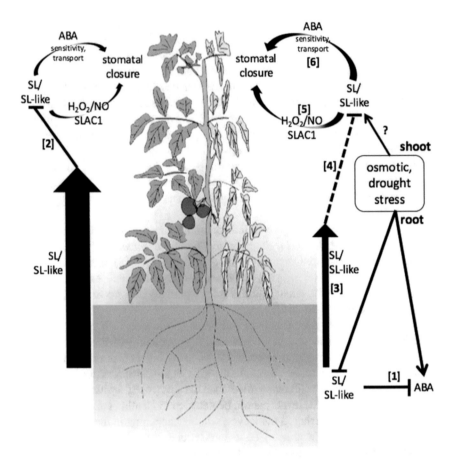

Fig. 2.5 A working model for SL action under osmotic stress in dicot plants. The main connections between SL (or SL-like molecules) and ABA in roots and shoots under drought stress are highlighted. SL may have a negative effect on osmotic stress-induced ABA levels in roots, suggesting that a drop in SL synthesis in this organ under osmotic stress may be required (but not sufficient) to let ABA levels rise [1]. The shootwards flow of SL molecules represses the transcription of SL/SL-like synthesis genes in shoots. This repression is stronger under normal conditions, when more SLs are produced in roots and translocated to the shoot [2] than under drought/osmotic stress (*see below*). SL synthesis is inhibited in roots under osmotic/drought stress, and, as a positive consequence for acclimatization, shootwards SL flow is decreased [3]. The transcription of SL synthesis genes is thus derepressed in shoots, likely increasing the SL levels [4] (dotted inhibition arrow indicates lower repression than in [2]). Shoot-produced SL molecules may induce SLAC1-dependent stomatal closure directly, by triggering the production of H_2O_2 and NO in guard cells [5]. They could also impact stomatal closure indirectly, by positively regulating ABA sensitivity or transport in guard cells [6]. Adapted with permission from: Cardinale et al. (2018) based on data by Liu et al. (2015), Li et al. (2017), Lv et al. (2018) and Visentin et al. (2016)

tolerant to drought (Haider et al. 2018). It may be that monocots and dicots exploit the regulatory mechanisms in a rather different fashion.

To further complicate the picture, it must be noted that if roots are colonized by AM fungi the effect of drought on SL production is reversed, and an increase in SLs occurs in roots (Aroca et al. 2013; Ruiz-Lozano et al. 2016). The benefit from this increase is yet unclear. It has been suggested that it may help to increase colonization of the roots by AM fungi, which supply water, and thus to alleviate drought stress (Lopez-Raez 2016). However if this was the underlying reason, SL synthesis should also increase in the absence of AM fungi, in order to increase the chances of recruiting symbiotic partners under drought, but this is not the case in most species. Clearly, more work is needed to disentangle the complexity of this system. Finally, it may be worth pointing out that, as different abiotic stresses (e.g. low nutrient versus drought) are associated to distinct changes in SL profiles in different organs, the changes in metabolite content displayed in response to combined stresses must be determined experimentally. For example, in *Lotus* roots, the SL decrease triggered by osmotic stress overrides the SL increase induced by P deprivation, when both stresses are applied together (Liu et al. 2015).

2.5 Interactions of SLs with Other Hormones

Over the past decade, it has been suggested that interactions occur between SLs and most other major plant hormones. This is often characterized as 'crosstalk', but this term has a rather specific definition, referring to interaction of signal transduction pathways. This is seldom the case in interactions between plant hormones. Instead, 'direct' interactions between plant hormones tend be 'cross-regulatory'; hormone A regulates the synthesis/transport/degradation/signalling of hormone B. There are also a plethora of 'indirect' interactions, where hormone A and B regulate the same process, or where long-term development responses to hormone A alter levels of hormone B. Here, we will focus on 'direct' interactions between SLs and other hormones. In doing so, it is important to consider whether a suggested interaction can be rationalized in terms of SL functionality. We can broadly characterize SLs as a systemically acting system that regulates responses to soil conditions, and it is through this prism that potential interactions between SLs and other hormones are best understood.

2.5.1 SLs and KL

Given the shared evolutionary history of SL and KL signalling, the molecular components shared between the pathways and the likely similarity of the ligands (Box 2.4), it would scarcely be surprising if there was true crosstalk between these pathways. For instance, it is easy to imagine KAI2 being able to target SMXL7 for

SCF^{MAX2}-mediated degradation, or D14 targeting SMAX1 for degradation. Indeed, in gymnosperms, there is no SMXL7 homologue, and it is plausible that both SL and KL signalling target SMAX1 for degradation (Walker and Bennett 2017). However, in angiosperms, there is very little indication of crosstalk between these pathways, nor is there clear evidence for other direct interactions. In the shoot, D14 and KAI2 activities seem to be completely separable, with D14 only targeting SMXL7 (and co-orthologues) and KAI2 only targeting SMAX1 (Soundappan et al. 2015). The majority of known SL effects can thus be explained by D14-mediated degradation of SMXL7. Even in tissues where both KAI2 and D14 are active, such as leaves, there is very little evidence for interaction between the pathways (Soundappan et al. 2015). In roots, the genetic evidence suggests that there could be crosstalk between the pathways. For instance, the *max2* phenotype of increased lateral root density (Sect. 2.3) seems to arise from combined lack of KAI2 and D14 signalling, but co-mutation of *SMXL6*, *SMXL7* and *SMXL8* completely suppresses the *max2* root phenotype (see also Sect. 2.3, Waters et al. 2017). This tentatively suggests that KAI2 might target SMXL6/SMXL7/SMXL8 for degradation in the root. However, since SMXL6/SMXL7/SMXL8 are completely epistatic to MAX2, this data is only *consistent* with the idea of crosstalk and not direct evidence of it. Ultimately, to test this idea, and possible SL/KL cross-regulation more broadly, more work will be needed.

2.5.2   SLs and Auxin

The best-characterized interactions are those between SL and auxin; indeed these regulatory interactions were already apparent even when the molecular identity of SLs was still unclear. There are direct interactions between SLs and auxin in both directions (Fig. 2.6). Auxin positively regulates expression of SL synthesis genes through its canonical signalling pathway, in both shoots and roots, in all species examined (Bainbridge et al. 2005; Foo et al. 2005; Johnson et al. 2006; Zou et al. 2006; Arite et al. 2007; Hayward et al. 2009). The consequences of this upregulation are not entirely clear. In heterozygous *bodenlos* auxin signalling mutants of *Arabidopsis*, *MAX3* and *MAX4* expression is strongly decreased. This appears to cause a high branching phenotype (similar to that of SL mutants themselves), which can be rescued by grafting to a wild-type (but not *max4*) root, or by *rac*-GR24 treatment (Hayward et al. 2009). However, SL mutant-like branching is not a general feature of auxin mutants, and expression of SL synthesis enzymes is rarely rate-limiting. Nevertheless, in the context of shoot branching, auxin upregulation of SL makes a great deal of sense. When active auxin-exporting shoots are present, upregulation of SL synthesis helps to repress further shoot branching and maintain the dominance of those shoots. Conversely, if auxin-exporting shoots are lost, SL synthesis declines, allowing easier activation of new branches.

In the other direction, SLs mainly appear to regulate the expression and activity of auxin transport proteins, particularly PIN-family auxin efflux carriers (Box 2.5). The

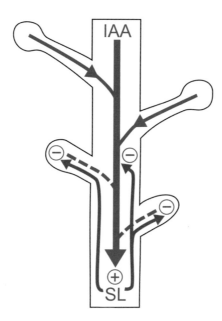

Fig. 2.6 SL-auxin cross-regulation. Auxin (IAA) is synthesized by actively growing shoot apices and transported through stems in the polar auxin transport stream (blue arrows). Auxin positively regulates synthesis of SLs in both the shoot and root. SL is transported shootwards (red arrows) and inhibits the activation of new shoot apices, reducing the total pool of auxin exported into the shoot system. SL also negatively regulates polar auxin transport through the shoot, which contributes to its inhibition of bud outgrowth. Thus SL action ultimately limits upregulation of SL synthesis

phenotypic significance of this regulation has been endlessly debated (Bennett et al. 2006; Brewer et al. 2009, 2015; Shinohara et al. 2013; Waters et al. 2017), but there is a wealth of evidence to suggest that it does occur. SLs seem able to regulate both the transcription of *PIN* genes and of PIN proteins dynamics in both shoot and root, though the exact effects are dependent on context and tissue. For instance, in the cambium and xylem parenchyma of stems, SLs promote endocytosis of PIN1 from the plasma membrane, in a fast and apparently transcriptionally independent manner (Crawford et al. 2010; Shinohara et al. 2013). In this way, SLs act to limit auxin transport through stems, and SL-deficient mutants have increased PIN protein levels and increased auxin transport (Beveridge et al. 2000; Bennett et al. 2006, 2016; Crawford et al. 2010; Shinohara et al. 2013), while mutants lacking SMXL6, SMXL7 and SMXL8 have decreased PIN levels and auxin transport (Soundappan et al. 2015; Liang et al. 2016). SLs may also have effects on transcription of *PIN* genes in the shoot, although this seems less consistent between tissues and species (Bennett et al. 2006; Young et al. 2014). By reducing auxin transport in the stem, SLs make it more difficult for shoot branches to export auxin. This appears to be one mechanism by which SLs regulate shoot branching, but it is likely that SLs also act directly in buds to repress branching as well (Brewer et al. 2015; Seale et al. 2017; Waters et al. 2017). Expression of SL synthesis genes is strongly upregulated in SL mutants, which might be caused by the increased auxin export from branches promoting expression of those genes. There is some evidence for this idea, but it is not currently conclusive (Hayward et al. 2009). Recent evidence suggests that SLs also regulate auxin synthesis gene expression in shoots over short timescales (30 min–1 h) possibly indicating direct regulation (Ligerot et al. 2017). This observation may explain the very high auxin levels observed in SL mutants (Prusinkiewicz et al. 2009).

In the root, SLs may also regulate auxin transport, though the evidence is less conclusive, partly because of the use of *max2* mutants and *rac*-GR24, which make it difficult to conclude the effects are due to SL and not KL signalling (Box 2.4). In the root meristem, long-term *rac*-GR24 treatment appears to affect PIN protein levels (Ruyter-Spira et al. 2011), but this appears to be a long-term developmental effect, rather than a direct effect of SL on PIN gene expression (Shinohara et al. 2013). In the epidermis, *rac*-GR24 seems to promote exocytosis of PIN2 to promote root hair elongation (Pandya-Kumar et al. 2014). Similarly, *rac*-GR24 seems to repress expression of PIN1 in the shootwards part of the root, which affects lateral root emergence in this zone (Jiang et al. 2016).

2.5.3   SLs and CKs

SLs and CKs are fundamentally antagonistic hormones. Both are synthesized in the root and convey information to the shoot about nutritional status, but they carry opposite messages. CKs generally relay positive information and promote shoot growth; SLs generally relay negative information and inhibit shoot growth. Thus, CK promotes branching, while SL represses it; SL promotes leaf senescence, while CK delays it. A reasonable expectation might therefore be that SL and CKs cross-regulate each other's synthesis, in order to strengthen their own effect on development. A variety of observations are consistent with this idea; for instance, pea SL mutants have significantly reduced CKs levels in xylem (Beveridge et al. 2000). Furthermore, branching in pea SL mutants is hypersensitive to CKs treatment, suggestive of cross-regulation. However, direct evidence of cross-regulation is limited, certainly for SL on CKs. The reduced CKs in SL mutants seem to arise from long-term developmental feedback, rather than from any effect of SL signalling on CKs biosynthesis itself (Young et al. 2014). Similarly, the hyper-sensitivity of SL mutants to CKs seems to arise from their antagonistic co-regulation of the *BRANCHED1* transcription factor in buds, rather than from any effect of SL on CKs signalling itself (Dun et al. 2012). In rice there is evidence that CKs treatment can downregulate SL synthesis genes over short timescales (Xu et al. 2015). However, the phenotypic consequences of this are unclear, because expression of SLs synthesis genes is rarely rate-limiting. Overall, since CKs synthesis in the roots is primarily responsive to soil nitrate levels, and SL synthesis generally responsive to phosphate levels, plants might use the CKs–SL ratio as an indicator of overall soil quality. This could explain the relative lack of cross-regulation between CKs and SL;

even though they are directly antagonistic in function, both signals are needed 'unmodified' in order for shoots to correctly gauge growth responses to soil resource availability.

2.5.4 SLs and GA

SLs and GAs overlap in certain developmental effects, such as leaf and internode elongation. There is also a clear analogy between the molecular mechanisms of SL and GA signalling, with α/β hydrolase proteins (D14 and GID1, respectively) triggering degradation of target proteins (SMXL7/D53 and DELLAs, respectively) in response to hormonal signals, mediated by SCF-type ubiquitin ligases (Wallner et al. 2016). It has thus been proposed that there may regulatory interactions between SL and GAs leading to their common effects. For instance, Nakamura et al. (2013) proposed that SL signalling targeted DELLA proteins for degradation, and Ito et al. (2017) proposed that GA regulates SL biosynthesis. However, currently, these findings have not been validated, and there is little evidence to support direct SLs and GAs interactions. For instance, there is very little phenotypic overlap between SL and GA mutants suggesting the hormones largely regulate separate processes (Bennett et al. 2016). Furthermore, in the case of stem elongation, the genetic evidence is clear that SLs and GAs act independently (de Saint Germain et al. 2013; Lantzouni et al. 2017). The direct effects of SLs and GAs on transcription in seedlings are almost entirely additive, again suggesting that these hormones act independently. Contrary to suggestions, SLs signalling also does not have any effect on DELLA stability (Bennett et al. 2016; Lantzouni et al. 2017). In rice, there does appear to be a long-term negative effect of GAs treatment on SLs biosynthesis gene expression and exudation of SLs from the root system (Ito et al. 2017). However, this only occurs after 24 h of GA treatment and is clearly not a direct response to GA treatment (Ito et al. 2017; Lantzouni et al. 2017). Overall, the functions of SLs and GAs in development are largely opposite, and in general, it would make little sense for them to interact in the manner previously proposed.

2.5.5 SLs and ABA

SLs and ABA have clear commonalities; both are carotenoid-derived signals and have overlapping roles in drought stress, particularly with respect to the impact of mycorrhization on water scarcity (Lopez-Raez 2016). Rather like the situation with GAs, this has prompted speculation about possible SLs and ABA interactions, but there is currently little hard evidence of direct interactions. The evidence for this is discussed above (Sect. 2.4). There are thus some indications of SL–ABA interactions may occur, but it is not clear that these are direct interactions between the hormones as opposed to more long-term homeostatic changes.

2.5.6   SLs and Brassinosteroids

A potential interaction between SLs and BR has been suggested by Wang et al. (2013), who proposed that BES1, a transcription factor that mediates BR responses, is a proteolytic target of SL/SCF^{MAX2} signalling. This argument is partly based on increased branching in the original *bes1-d* mutant in which BES1 is stabilized (Wang et al. 2013). However, reanalysis of a backcrossed *bes1* mutant lines suggests that BES1 activity has no relevance for SL phenotypic effects and does not confer SL insensitivity (Bennett et al. 2016). Thus, although it cannot be ruled out that SL signalling (or indeed KL signalling) does alter the stability of BES1, if this does occur, it seems to have little relevance for the well-defined effects of SLs on plant development.

2.5.7   Summary

Although many interactions have been proposed between SLs and other hormones, most of these do not represent true regulatory interactions, but rather more indirect processes. This includes joint regulation of downstream targets by SLs and other hormones, and changes in SL homeostasis caused by altered development downstream of other hormones. Other proposed interactions have not been successfully validated. Indeed, only in the case of auxin–SL interactions is there a solid body of evidence for meaningful direct interactions. It is not a coincidence that those auxin–SL interactions are the interactions that are most easily rationalized as part of SL functionality.

2.6   Future Outlook

SLs are now universally accepted as a new class of phytohormones. In spite of the growing knowledge about their structure and molecular mechanisms of action, there are still many open questions about their role as plant hormones. First, it is of major interest to discern the functional specificity of SL signalling and its potential crosstalk with KL signalling during plant development and acclimation responses. Second, it is still not fully understood how the systemic action of SL signalling regulates shoot and root development, and whether these two processes are interconnected. Third, the molecular basis of the divergent roles of SLs across angiosperms remains to be established. Nevertheless, the impact of SLs on key developmental processes such as plant architecture, and their involvement in the acclimation of plants to environmental stresses, raises the possibility of using these hormones and signalling pathways as agricultural tools to optimize crop plant architecture and resilience to abiotic stress.

2.7 Conclusions

- SLs are a new class of plant hormones.
- SLs control aerial and underground architecture.
- Some SL functions are widely conserved across angiosperms; others seem to be species-specific or detectable only under stress conditions.
- Under nutrient (P or N) deprivation, SLs help the plant improve mineral uptake and optimize resource allocation.
- SLs, probably in coordination with KL, contribute to the plant acclimation responses to water stress.
- Direct regulatory interactions occur between SL and auxin, which determine shoot branching patterns and perhaps also root architecture.

Glossary

Abscisic acid (ABA) Carotenoid-derived phytohormone that regulates many aspects of plant growth, development and cellular signalling. ABA controls seed dormancy, seed maturation, vegetative growth and responses to various environmental stimuli such as stomatal closure during drought.

Acclimation Relatively fast and reversible changes that individual organisms undergo in response to environmental changes.

Adaptation Evolutionary process that affects species (or groups of individuals) and leads to better fitting to the habitat via genetic, physical and chemical adjustments.

Amyloplasts Organelles found in some plant cells responsible for the synthesis and storage of starch granules, through the polymerization of glucose. Sedimentation of amyloplasts is associated with gravity perception in specialized gravity-sensing cells.

Anthocyanins Plant pigments of the flavonoid family synthesized via the phenylpropanoid pathway. They are found in all tissues of higher plants. Depending on cellular pH, anthocyanins may appear red, purple or blue.

Apical dominance Phenomenon whereby actively growing 'dominant' shoot branches prevent the outgrowth of new branches. Removal of the apices of dominant shoots allows the outgrowth of previously inhibited branches.

Arbuscular mycorrhiza Type of mycorrhiza in which the symbiotic fungus (AM fungi or AMF) penetrates into plant cells and forms characteristic exchange bodies known as arbuscles.

Auxins Phytohormone with a huge array of roles in the coordination of plant growth and development. Endogenous auxin is indole-3-acetic acid, but a range of additional auxin-like molecules and synthetic auxin analogues with related structures also exist.

Brassinosteroids Polyhydroxysteroid phytohormones. They promote stem elongation, cell division, root development and stress responses. Brassinolide is the most common brassinosteroid.

Cambium Vascular-associated tissue layer that can undergo divisions to allow the radial expansion of plant tissues ('secondary growth').

Crosstalk Interaction of signal transduction pathways.

Cytokinins (CKs) Class of phytohormones with a range of structural forms. Cytokinins act as both root-to-shoot and shoot-to-root signals and are involved in various aspects of development including promoting shoot branching, inhibiting leaf senescence and promoting meristematic activity.

Ethylene Gaseous hydrocarbon with the formula $H_2C{=}CH_2$. It is a natural phytohormone that stimulates fruit ripening, flower opening and abscission (or shedding) of leaves. It is also used in agriculture to promote ripening of certain fruits.

Flowering time The time taken for a flowering plant to pass through the vegetative phase and enter the reproductive phase. Typically measured either in absolute time or as number of nodes a plant produces before formation of the first flower.

Gibberellic acid (GA) Phytohormone involved in breaking seed dormancy, promoting seed development, stimulating stem and root growth, inducing mitotic division in the leaves of some plants and promoting vegetative and floral growth.

Gravitropism Process of differential growth in response to gravity. It is a general feature of all plants. Roots show positive gravitropism (i.e. they grow in the direction of gravitational vector, i.e. downwards), while stems show negative gravitropism.

Karrikins Butenolide compounds found in the smoke of burnt plant material, which act as plant growth regulators and stimulate the germination of seeds. Karrikins act through the KAI2 receptor but are not an endogenous ligand for KAI2 (KAI2-ligand, KL).

Meristems Specialized areas of tissue in which the vast majority of cell divisions in plant occur. Meristems generate new cells that allow the growth of the plant in various dimensions. In flowering plants, the embryonic root and shoot apical meristems give rise to the entire root and shoot system, respectively. Axillary meristems are secondary shoot meristems formed in the leaf axils. Cambial meristems are responsible for the radial growth and thickening of the stem.

Mycorrhiza Symbiotic association between a fungus and a plant. Mycorrhizas play important roles in soil biology and soil chemistry. Mycorrhizas may involve colonization of the extracellular space (ectomycorrhizas) or intracellular colonization (arbuscular mycorrhizas). The association is generally mutualistic, but in particular species or in particular circumstances, either partner can parasitize the other.

Nitrogen fixation Nitrogen fixation is a process by which atmospheric nitrogen (N2) is converted into ammonia or other organic molecules. Nitrogen fixation is carried out naturally in the soil by nitrogen-fixing bacteria. Certain nitrogen-fixing bacteria have symbiotic relationships with plant groups. Especially notable is the association between legumes (Fabaceae) and *Rhizobia* spp.

Nodulation In legumes, root nodules are specialized structures that host the symbiotic nitrogen-fixing bacteria. They are typically formed under nitrogen-limited conditions.

Osmotic pressure Minimum pressure required to be applied to a solution to prevent the inward flow of its pure solvent across a semipermeable membrane. It is also defined as the measure of the tendency of a solution to take in pure solvent by osmosis. This process is of vital importance in biology as the cell's membrane is semipermeable.

Plant phenotypic plasticity The ability of a plant genotype to generate different phenotypes in response to varying environmental conditions.

Phloem Vascular plant tissue that transports the soluble organic compounds made during photosynthesis, photosynthates, in particular sucrose, to parts of the plant where needed.

Racemic-**GR24** (*rac*-**GR24**) One of the most commonly used synthetic strigolactone analogues.

Rhizobia Bacteria that fix nitrogen after becoming established inside root nodules of legumes (Fabaceae).

Rhizosphere The soil surrounding and directly influenced by plant roots.

RNA interference (RNAi) Biological process in which RNA molecules inhibit gene expression or translation, by neutralizing targeted mRNA molecules. The generation of RNAi transgenic lines is an approach commonly used in plant research to cause precise and efficient gene suppression (knock-down).

Root exudates Chemicals that are exported by the roots into the rhizosphere, which play a variety of roles in communication with microorganisms and manipulation of the physical and chemical properties of the soil.

Senescence A process of deliberate organ breakdown, allowing recycling of nutrients to growing and storage organs of the plant. Senescence typically occurs in older organs, to fuel the development of new organs. Senescence may be increased under stress conditions.

Stolons Stems, often called runners, which grow at the soil surface or just below ground that form adventitious roots at the nodes and new plants from the buds. They support vegetative propagation.

Stomata Pores found in the plant epidermis that facilitate gas exchange. They are bordered by a pair of specialized epidermal cells known as guard cells that are responsible for regulating the size of the stomatal opening.

Stomatal conductance It is the measure of the rate of passage of gases through the stomata of a leaf, mostly carbon dioxide entering, and water vapour exiting through the stomata of a leaf. It is directly related to the absolute concentration gradient of water vapour from the leaf to the atmosphere.

Tuberization The process by which some plant species develop tubers, enlarged modified stems used as underground storage organs for nutrients.

Vasculature Continuous tissue system that allow transport of water and nutrients around the plant body. There are two main types of vascular element, xylem and phloem. Vascular elements typically occur in 'vascular bundles', which also include supporting and protective tissues.

Xylem One of the two types of transport tissue in vascular plants. Its main function is to transport water and mineral nutrients from roots to shoots and leaves.

Water potential Potential energy of water per unit volume relative to pure water in reference conditions.

References

Aguilar-Martínez JA, Poza-Carrión C, Cubas P (2007) *Arabidopsis BRANCHED1* acts as an integrator of branching signals within axillary buds. Plant Cell 19:458–472

Akiyama K, Matsuzaki K, Hayashi H (2005) Plant sesquiterpenes induce hyphal branching in arbuscular mycorrhizal fungi. Nature 435:824–827

Agusti J, Herold S, Schwarz M, Sanchez P, Ljung K, Dun EA, Brewer PB, Beveridge CA, Sieberer T, Sehr EM et al (2011) Strigolactone signalling is required for auxin-dependent stimulation of secondary growth in plants. Proc Natl Acad Sci U S A 108:20242–20247

Andreo-Jimenez B, Ruyter-Spira C, Bouwmeester HJ, Lopez-Raez JA (2015) Ecological relevance of strigolactones in nutrient uptake and other abiotic stresses, and in plant-microbe interactions below-ground. Plant Soil 394:1–19

Arite T, Umehara M, Ishikawa S, Hanada A, Maekawa M, Yamaguchi S, Kyozuka J (2009) *d14*, a strigolactone-insensitive mutant of rice, shows an accelerated outgrowth of tillers. Plant Cell Physiol 50:1416–1424

Arite T, Iwata H, Ohshima K, Maekawa M, Nakajima M, Kojima M, Sakakibara H, Kyozuka J (2007) DWARF10, an RMS1/MAX4/DAD1 ortholog, controls lateral bud outgrowth in rice. Plant J 51:1019–1029

Aroca R, Ruiz-Lozano JM, Zamarreno AM, Paz JA, Garcia-Mina JM, Pozo MJ, López-Ráez JA (2013) Arbuscular mycorrhizal symbiosis influences strigolactone production under salinity and alleviates salt stress in lettuce plants. J Plant Physiol 170:47–55

Artuso E, Ghibaudi E, Lace B, Marabello D, Vinciguerra D, Lombardi C, Koltai H, Kapulnik Y, Novero M, Occhiato EG, Scarpi D, Parisotto S, Deagostino A, Venturello P, Mayzlish-Gati E, Bier A, Prandi C (2015) Stereochemical assignment of strigolactone analogues confirms their selective biological activity. J Nat Prod 78:2624–2633

Bainbridge K, Sorefan K, Ward S, Leyser O (2005) Hormonally controlled expression of the *Arabidopsis* MAX4 shoot branching regulatory gene. Plant J 44:569–580

Balzergue C, Puech-Pagès V, Bécard G, Rochange SF (2011) The regulation of arbuscular mycorrhizal symbiosis by phosphate in pea involves early and systemic signalling events. J Exp Bot 62:1049–1060

Basu S, Ramegowda V, Kumar A, Pereira A (2016) Plant adaptation to drought stress. F1000 Res 5:1554

Bennett T, Sieberer T, Willett B, Booker J, Luschnig C, Leyser O (2006) The *Arabidopsis* MAX pathway controls shoot branching by regulating auxin transport. Curr Biol 16:553–563

Bennett T, Liang Y, Seale M, Ward S, Müller D, Leyser O (2016) Strigolactone regulates shoot development through a core signalling pathway. Biol Open 5:1806–1820

Beveridge CA, Dun EA, Rameau C (2009) Pea has its tendrils in branching discoveries spanning a century from auxin to strigolactones. Plant Physiol 151:985–990

Beveridge CA, Symons GM, Turnbull CG (2000) Auxin inhibition of decapitation-induced branching is dependent on graft-transmissible signals regulated by genes *Rms1* and *Rms2*. Plant Physiol 123:689–698

Bodner G, Nakhforoosh A, Kaul HP (2015) Management of crop water under drought: a review. Agron Sustain Dev 35:401–442

Booker J, Auldridge M, Wills S, McCarty D, Klee H, Leyser O (2004) MAX3/CCD7 is a carotenoid cleavage dioxygenase required for the synthesis of a novel plant signalling molecule. Curr Biol 14:1232–1238

Braun N, de Saint Germain A, Pillot J-PJ-P, Boutet-Mercey S, Dalmais M, Antoniadi I, Li X, Maia-Grondard A, Le Signor C, Bouteiller N et al (2012) The pea TCP transcription factor PsBRC1 acts downstream of strigolactones to control shoot branching. Plant Physiol 158:225–238

Breuillin F, Schramm J, Hajirezaei M, Ahkami A, Favre P, Druege U, Hause B, Bucher M, Kretzschmar T, Bossolini E, Kuhlemeier C, Martinoia E, Franken P, Scholz U, Reinhardt D (2010) Phosphate systemically inhibits development of arbuscular mycorrhiza in *Petunia hybrida* and represses genes involved in mycorrhizal functioning. Plant J 64:1002–1017

Brewer PB, Dun EA, Gui R, Mason MG, Beveridge CA (2015) Strigolactone inhibition of branching independent of polar auxin transport. Plant Physiol 168:1820–1829

Brewer PB, Dun EA, Ferguson BJ, Rameau C, Beveridge CA (2009) Strigolactone acts downstream of auxin to regulate bud outgrowth in pea and *Arabidopsis*. Plant Physiol 150:482–493

Bu Q, Lv T, Shen H, Luong P, Wang J, Wang Z, Huang Z, Xiao L, Engineer C, Kim TH, Schroeder JI, Huq E (2014) Regulation of drought tolerance by the F-box protein MAX2 in *Arabidopsis*. Plant Physiol 164:424–439

Bythell-Douglas R, Rothfels CJ, Stevenson DWD, Graham SW, Wong GK, Nelson DC, Bennett T (2017) Evolution of strigolactone receptors by gradual neo-functionalization of KAI2 paralogues. BMC Biol 15:52

Cardinale F, Korwin Krukowski P, Schubert A, Visentin I (2018) Strigolactones: mediators of osmotic stress responses with a potential for agrochemical manipulation of crop resilience. J Exp Bot 69:2291–2303

Chaves MM, Maroco JP, Pereira JS (2003) Understanding plant responses to drought-from genes to the whole plant. Funct Plant Biol 30:239–264

Chevalier F, Nieminen K, Sánchez-Ferrero JC, Rodríguez ML, Chagoyen M, Hardtke CS, Cubas P (2014) Strigolactone promotes degradation of DWARF14, an α/β hydrolase essential for strigolactone signaling in *Arabidopsis*. Plant Cell 26:1134–1150

Conn CE, Nelson DC (2016) Evidence that KARRIKIN-INSENSITIVE2 (KAI2) receptors may perceive an unknown signal that is not karrikin or strigolactone. Front Plant Sci 6:1219

Cook CE, Whichard LP, Turner B, Wall ME, Egley GH (1966) Germination of witchweed (*Striga lutea* Lour.): isolation and properties of a potent stimulant. Science 154:1189–1190

Crawford S, Shinohara N, Sieberer T, Williamson L, George G, Hepworth J, Müller D, Domagalska MA, Leyser O (2010) Strigolactones enhance competition between shoot branches by dampening auxin transport. Development 137:2905–2913

Czarnecki O, Yang J, Weston DJ, Tuskan GA, Chen JG (2013) A dual role of strigolactones in phosphate acquisition and utilization in plants. Int J Mol Sci 14:7681–7701

De Cuyper C, Fromentin J, Yocgo RE, De Keyser A, Guillotin B, Kunert K, Boyer FD, Goormachtig S (2015) From lateral root density to nodule number, the strigolactone analogue GR24 shapes the root architecture of *Medicago truncatula*. J Exp Bot 66:137–146

de Jong M, George G, Ongaro V, Williamson L, Willetts B, Ljung K, McCulloch H, Leyser O (2014) Auxin and strigolactone signalling are required for modulation of Arabidopsis shoot branching by nitrogen supply. Plant Physiol 166:384–395

de Saint Germain A, Ligerot Y, Dun EA, Pillot J-PJ-P, Ross JJ, Beveridge CA, Rameau C (2013) Strigolactones stimulate internode elongation independently of gibberellins. Plant Physiol 163:1012–1025

Doebley J, Stec A, Hubbard L (1997) The evolution of apical dominance in maize. Nature 386:485–488

Dun EA, de Saint Germain A, Rameau C, Beveridge CA (2012) Antagonistic action of strigolactone and cytokinin in bud outgrowth control. Plant Physiol 158:487–498

Foo E, Bullier E, Goussot M, Foucher F, Rameau C, Beveridge CA (2005) The branching gene *RAMOSUS1* mediates interactions among two novel signals and auxin in pea. Plant Cell 17:464–474

Gomez-Roldan V, Fermas S, Brewer PB, Puech-Pagès V, Dun EA, Pillot J-P, Letisse F, Matusova R, Danoun S, Portais JC et al (2008) Strigolactone inhibition of shoot branching. Nature 455:189–194

Guan JC, Koch KE, Suzuki M, Wu S, Latshaw S, Petruff T, Goulet C, Klee HJ, McCarty DR (2012) Diverse roles of strigolactone signalling in maize architecture and the uncoupling of a branching-specific subnetwork. Plant Physiol 160:1303–1317

Ha CV, Leyva-González MA, Osakabe Y, Tran UT, Nishiyama R, Watanabe Y, Tanaka M, Seki M, Yamaguchi S, Dong NV, Yamaguchi-Shinozaki K, Shinozaki K, Herrera-Estrella L, Tran LS (2014) Positive regulatory role of strigolactone in plant responses to drought and salt stress. Proc Natl Acad Sci U S A 111:851–856

Haider I, Andreo-Jimenez B, Bruno M, Bimbo A, Floková K, Abuauf H, Otang Ntui V, Guo X, Charnikhova T, Al-Babili S, Bouwmeester HJ, Ruyter-Spira C (2018) The interaction of strigolactones with abscisic acid during the drought response in rice. J Exp Bot 69:2403–2414

Hamiaux C, Drummond RS, Janssen BJ, Ledger SE, Cooney JM, Newcomb RD, Snowden KC (2012) DAD2 is an α/β hydrolase likely to be involved in the perception of the plant branching hormone, strigolactone. Curr Biol 22:2032–2036

Hayward A, Stirnberg P, Beveridge C, Leyser O (2009) Interactions between auxin and strigolactone in shoot branching control. Plant Physiol 151:400–412

Hu Q, He Y, Wang L, Liu S, Meng X, Liu G, Jing Y, Chen M, Song X, Jiang L, Yu H, Wang B, Li J (2017) DWARF14, a receptor covalently linked with the active form of strigolactones, undergoes strigolactone-dependent degradation in rice. Front Plant Sci 8:1935

Ito S, Yamagami D, Umehara M, Hanada A, Yoshida S, Sasaki Y, Yajima S, Kyozuka J, Ueguchi-Tanaka M, Matsuoka M, Shirasu K, Yamaguchi S, Asami T (2017) Regulation of strigolactone biosynthesis by gibberellin signalling. Plant Physiol 174:1250–1259

Ito S, Ito K, Abeta N, Takahashi R, Sasaki Y, Yajima S (2016) Effects of strigolactone signalling on Arabidopsis growth under nitrogen deficient stress condition. Plant Signal Behav 11:e1126031

Ito S, Nozoye T, Sasaki E, Imai M, Shiwa Y, Shibata-Hatta M, Ishige T, Fukui K, Ito K, Nakanishi H, Nishizawa NK, Yajima S, Asami T (2015) Strigolactone regulates anthocyanin accumulation, acid phosphatases production and plant growth under low phosphate condition in Arabidopsis. PLoS One 10:e0119724

Jiang L, Liu X, Xiong G, Liu H, Chen F, Wang L, Meng X, Liu G, Yu H, Yuan Y, Yi W, Zhao L, Ma H, He Y, Wu Z, Melcher K, Qian Q, Xu HE, Wang Y, Li J (2013) DWARF 53 acts as a repressor of strigolactone signalling in rice. Nature 504:401–405

Jiang L, Matthys C, Marquez-Garcia B, De Cuyper C, Smet L, De Keyser A, Boyer FD, Beeckman T, Depuydt S, Goormachtig S (2016) Strigolactones spatially influence lateral root development through the cytokinin signalling network. J Exp Bot 67:79–89

Johnson X, Brcich T, Dun EA, Goussot M, Haurogné K, Beveridge CA, Rameau C (2006) Branching genes are conserved across species. Genes controlling a novel signal in pea are coregulated by other long-distance signals. Plant Physiol 142:1014–1026

Kapulnik Y, Delaux PM, Resnick N, Mayzlish-Gati E, Wininger S, Bhattacharya C, Séjalon-Delmas N, Combier JP, Bécard G, Belausov E, Beeckman T, Dor E, Hershenhorn J, Koltai H (2011a) Strigolactones affect lateral root formation and root-hair elongation in Arabidopsis. Planta 233:209–216

Kapulnik Y, Resnick N, Mayzlish-Gati E, Kaplan Y, Wininger S, Hershenhorn J, Koltai H (2011b) Strigolactones interact with ethylene and auxin in regulating root-hair elongation in Arabidopsis. J Exp Bot 62:2915–2924

Kohlen W, Charnikhova T, Liu Q, Bours R, Domagalska MA, Beguerie S, Verstappen F, Leyser O, Bouwmeester H, Ruyter-Spira C (2011) Strigolactones are transported through the xylem and play a key role in shoot architectural response to phosphate deficiency in nonarbuscular mycorrhizal host Arabidopsis. Plant Physiol 155:974–987

Kohlen W, Charnikhova T, Lammers M, Pollina T, Tóth P, Haider I, Pozo MJ, de Maagd RA, Ruyter-Spira C, Bouwmeester HJ, López-Ráez JA (2012) The tomato CAROTENOID CLEAVAGE DIOXYGENASE8 (SlCCD8) regulates rhizosphere signaling, plant architecture and affects reproductive development through strigolactone biosynthesis. New Phytol 196:535–547

Koltai H, Dor E, Hershenhorn J, Joel DM, Weininger S, Lekalla S, Shealtiel H, Bhattacharya C, Eliahu E, Resnick N, Barg R, Kapulnik Y (2010) Strigolactones' effect on root growth and root-hair elongation may be mediated by auxin-efflux carriers. J Plant Growth Regul 29:129–136

Lambers H, Raven JA, Shaver GR, Smith SE (2008) Plant nutrient-acquisition strategies change with soil age. Trends Ecol Evol 23:95–103

Lanfranco L, Fiorilli V, Venice F, Bonfante P (2018) Strigolactones cross the kingdoms: plants, fungi, and bacteria in the arbuscular mycorrhizal symbiosis. J Exp Bot 69:2175–2188

Lantzouni O, Klermund C, Schwechheimer C (2017) Largely additive effects of gibberellin and strigolactone on gene expression in *Arabidopsis thaliana* seedlings. Plant J 92:924–938

Lauressergues D, André O, Peng J, Wen J, Chen R, Ratet P, Tadege M, Mysore KS, Rochange SF (2015) Strigolactones contribute to shoot elongation and to the formation of leaf margin serrations in *Medicago truncatula* R108. J Exp Bot 66:1237–1244

Li S, Chen L, Li Y, Yao R, Wang F, Yang M, Gu M, Nan F, Xie D, Yan J (2016) Effect of GR24 stereoisomers on plant development in *Arabidopsis*. Mol Plant 9:1432–1435

Li W, Nguyen KH, Chu HD, Ha CV, Watanabe Y, Osakabe Y, Leyva-González MA, Sato M, Toyooka K, Voges L, Tanaka M, Mostofa MG, Seki M, Seo M, Yamaguchi S, Nelson DC, Tian C, Herrera-Estrella L, Tran LP (2017) The karrikin receptor KAI2 promotes drought resistance in *Arabidopsis thaliana*. PLoS Genet 13:e1007076

Liang Y, Ward S, Li P, Bennett T, Leyser O (2016) SMAX1-LIKE7 signals from the nucleus to regulate shoot development in *Arabidopsis* via partially EAR motif-independent mechanisms. Plant Cell 28:1581–1601

Ligerot Y, de Saint Germain A, Waldie T, Troadec C, Citerne S, Kadakia N, Pillot JP, Prigge M, Aubert G, Bendahmane A, Leyser O, Estelle M, Debellé F, Rameau C (2017) The pea branching *RMS2* gene encodes the PsAFB4/5 auxin receptor and is involved in an auxin-strigolactone regulation loop. PLoS Genet 13:e1007089

Linkohr BI, Williamson LC, Fitter AH, Leyser HM (2002) Nitrate and phosphate availability and distribution have different effects on root system architecture of *Arabidopsis*. Plant J 29:751–760

Liu J, Novero M, Charnikhova T, Ferrandino A, Schubert A, Ruyter-Spira C, Bonfante P, Lovisolo C, Bouwmeester HJ, Cardinale F (2013) CAROTENOID CLEAVAGE DIOXYGENASE 7 modulates plant growth, reproduction, senescence, and determinate nodulation in the model legume *Lotus japonicus*. J Exp Bot 64:1967–1981

Liu J, He H, Vitali M, Visentin I, Charnikhova T, Haider I, Schubert A, Ruyter-Spira C, Bouwmeester HJ, Lovisolo C, Cardinale F (2015) Osmotic stress represses strigolactone biosynthesis in *Lotus japonicus* roots: exploring the interaction between strigolactones and ABA under abiotic stress. Planta 241:1435–1451

Lopez-Raez JA (2016) How drought and salinity affect arbuscular mycorrhizal symbiosis and strigolactone biosynthesis? Planta 243:1375–1385

Lopez-Raez JA, Shirasu K, Foo E (2017) Strigolactones in plant interactions with beneficial and detrimental organisms: the Yin and Yang. Trends Plant Sci 22:527–537

Luo L, Wang H, Liu X, Hu J, Zhu X, Pan S, Qin R, Wang Y, Zhao P, Fan X, Xu G (2018) Strigolactones affect the translocation of nitrogen in rice. Plant Sci 270:190–197

Lv S, Zhang Y, Li C, Liu Z, Yang N, Pan L, Wu J, Wang J, Yang J, Lv Y, Zhang Y, Jiang W, She X, Wang G (2018) Strigolactone-triggered stomatal closure requires hydrogen peroxide synthesis and nitric oxide production in an abscisic acid-independent manner. New Phytol 217:290–304

Lynch JP, Brown KM (2001) Topsoil foraging—an architectural adaptation of plants to low phosphorus availability. Plant Soil 237:225–237

Madmon O, Mazuz M, Kumari P, Dam A, Ion A, Mayzlish-Gati E, Belausov E, Wininger S, Abu-Abied M, McErlean CS, Bromhead LJ, Perl-Treves R, Prandi C, Kapulnik Y, Koltai H (2016) Expression of *MAX2* under SCARECROW promoter enhances the strigolactone/MAX2 dependent response of *Arabidopsis* roots to low-phosphate conditions. Planta 243:1419–1427

Martín-Trillo M, González-Grandío EG, Serra F, Marcel F, Rodríguez-Buey ML, Schmitz G, Theres K, Bendahmane A, Dopazo H, Cubas P (2011) Role of tomato *BRANCHED1*-like genes in the control of shoot branching. Plant J 67:701–714

Mashiguchi K, Sasaki E, Shimada Y, Nagae M, Ueno K, Nakano T, Yoneyama K, Suzuki Y, Asami T (2009) Feedback-regulation of strigolactone biosynthetic genes and strigolactone-related genes in Arabidopsis. Biosci Biotechnol Biochem 73:2460–2465

Mason MG, Ross JJ, Babst BA, Wienclaw BN, Beveridge CA (2014) Sugar demand, not auxin, is the initial regulator of apical dominance. Proc Natl Acad Sci U S A 111:6092–6097

Matthys C, Walton A, Struk S, Stes E, Boyer FD, Gevaert K, Goormachtig S (2016) The whats, the wheres and the hows of strigolactone action in the roots. Planta 243:1327–1337

Matusova R, Rani K, Verstappen FWA, Franssen MCR, Beale MH, Bouwmeester HJ (2005) The strigolactone germination stimulants of the plant-parasitic *Striga* and *Orobanche* spp. are derived from the carotenoid pathway. Plant Physiol 139:920–934

Mayzlish-Gati E, De-Cuyper C, Goormachtig S, Beeckman T, Vuylsteke M, Brewer PB, Beveridge CA, Yermiyahu U, Kaplan Y, Enzer Y, Wininger S, Resnick N, Cohen M, Kapulnik Y, Koltai H (2012) Strigolactones are involved in root response to low phosphate conditions in *Arabidopsis*. Plant Physiol 160:1329–1341

Minakuchi K, Kameoka H, Yasuno N, Umehara M, Luo L, Kobayashi K, Hanada A, Ueno K, Asami T, Yamaguchi S et al (2010) FINE CULM1 (FC1) works downstream of strigolactones to inhibit the outgrowth of axillary buds in rice. Plant Cell Physiol 51:1127–1135

Mostofa MG, Li W, Nguyen KH, Fujita M, Tran LP (2018) Strigolactones in plant adaptation to abiotic stresses: an emerging avenue of plant research. Plant Cell Environ 41(10):2227–2243

Nakamura H, Xue YL, Miyakawa T, Hou F, Qin HM, Fukui K, Shi X, Ito E, Ito S, Park SH, Miyauchi Y, Asano A, Totsuka N, Ueda T, Tanokura M, Asami T (2013) Molecular mechanism of strigolactone perception by DWARF14. Nat Commun 4:2613

Nelson DC, Scaffidi A, Dun EA, Waters MT, Flematti GR, Dixon KW, Beveridge CA, Ghisalberti EL, Smith SM (2011) F-box protein MAX2 has dual roles in karrikin and strigolactone signalling in *Arabidopsis thaliana*. Proc Natl Acad Sci U S A 108:8897–8902

Pandya-Kumar N, Shema R, Kumar M, Mayzlish-Gati E, Levy D, Zemach H, Belausov E, Wininger S, Abu-Abied M, Kapulnik Y, Koltai H (2014) Strigolactone analog GR24 triggers changes in PIN2 polarity, vesicle trafficking and actin filament architecture. New Phytol 202:1184–1196

Parniske M (2008) Arbuscular mycorrhiza: the mother of plant root endosymbioses. Nat Rev Microbiol 6:763–775

Pasare SA, Ducreux LJM, Morris WL, Campbell R, Sharma SK, Roumeliotis E, Kohlen W, van der Krol S, Bramley PM, Roberts AG et al (2013) The role of the potato (*Solanum tuberosum*) *CCD8* gene in stolon and tuber development. New Phytol 198:1108–1120

Prusinkiewicz P, Crawford S, Smith RS, Ljung K, Bennett T, Ongaro V, Leyser O (2009) Control of bud activation by an auxin transport switch. Proc Natl Acad Sci U S A 106:17431–17436

Rasmussen A, Mason MG, De Cuyper C, Brewer PB, Herold S, Agusti J, Geelen D, Greb T, Goormachtig S, Beeckman T, Beveridge CA (2012) Strigolactones suppress adventitious rooting in *Arabidopsis* and pea. Plant Physiol 158:1976–1987

Ruiz-Lozano JM, Aroca R, Zamarreno AM, Molina S, Andreo-Jimenez B, Porcel R, Garcia-Mina JM, Ruyter-Spira C, Lopez-Raez JA (2016) Arbuscular mycorrhizal symbiosis induces strigolactone biosynthesis under drought and improves drought tolerance in lettuce and tomato. Plant Cell Environ 39:441–452

Ruyter-Spira C, Kohlen W, Charnikhova T, van Zeijl A, van Bezouwen L, de Ruijter N, Cardoso C, Lopez-Raez JA, Matusova R, Bours R, Verstappen F, Bouwmeester H (2011) Physiological effects of the synthetic strigolactone analog GR24 on root system architecture in *Arabidopsis*: another belowground role for strigolactones? Plant Physiol 155:721–734

Sairam RK, Tyagi A (2004) Physiology and molecular biology of salinity stress tolerance in plants. Curr Sci 86:407–421

Sang D, Chen D, Liu G, Liang Y, Huang L, Meng X, Chu J, Sun X, Dong G, Yuan Y, Qian Q, Li J, Wang Y (2014) Strigolactones regulate rice tiller angle by attenuating shoot gravitropism through inhibiting auxin biosynthesis. Proc Natl Acad Sci U S A 111:11199–11204

Scaffidi A, Waters MT, Ghisalberti EL, Dixon KW, Flematti GR, Smith SM (2013) Carlactone-independent seedling morphogenesis in *Arabidopsis*. Plant J 76:1–9

Scaffidi A, Waters MT, Sun YK, Skelton BW, Dixon KW, Ghisalberti EL, Flematti GR, Smith SM (2014) Strigolactone hormones and their stereoisomers signal through two related receptor proteins to induce different physiological responses in *Arabidopsis*. Plant Physiol 165:1221–1232

Seale M, Bennett T, Leyser O (2017) *BRC1* expression regulates bud activation potential but is not necessary or sufficient for bud growth inhibition in *Arabidopsis*. Development 144:1661–1673

Shinohara N, Taylor C, Leyser O (2013) Strigolactone can promote or inhibit shoot branching by triggering rapid depletion of the auxin efflux protein PIN1 from the plasma membrane. PLoS Biol 11:e1001474

Simons JL, Napoli CA, Janssen BJ, Plummer KM, Snowden KC (2006) Analysis of the *DECREASED APICAL DOMINANCE* genes of petunia in the control of axillary branching. Plant Physiol 143:697–706

Snowden KC, Simkin AJ, Janssen BJ, Templeton KR, Loucas HM, Simons JL, Karunairetnam S, Gleave AP, Clark DG, Klee HJ (2005) The *decreased apical dominance1/Petunia hybrida CAROTENOID CLEAVAGE DIOXYGENASE8* gene affects branch production and plays a role in leaf senescence, root growth, and flower development. Plant Cell 17:746–759

Sorefan K, Booker J, Haurogné K, Goussot M, Bainbridge K, Foo E, Chatfield S, Ward S, Beveridge C, Rameau C, Leyser O (2003) *MAX4* and *RMS1* are orthologous dioxygenase-like genes that regulate shoot branching in *Arabidopsis* and pea. Genes Dev 17:1469–1474

Soundappan I, Bennett T, Morffy N, Liang Y, Stanga JP, Abbas A, Leyser O, Nelson D (2015) SMAX1-LIKE/D53 family members enable distinct MAX2-dependent responses to strigolactones and karrikins in *Arabidopsis*. Plant Cell 27:3143–3159

Stirnberg P, Furner IJ, Leyser OHM (2007) MAX2 participates in an SCF complex which acts locally at the node to suppress shoot branching. Plant J 50:80–94

Stirnberg P, van De Sande K, Leyser OHM (2002) MAX1 and MAX2 control shoot lateral branching in *Arabidopsis*. Development 129:1131–1141

Sun H, Tao J, Liu S, Huang S, Chen S, Xie X, Yoneyama K, Zhang Y, Xu G (2014) Strigolactones are involved in phosphate- and nitrate-deficiency-induced root development and auxin transport in rice. J Exp Bot 65:6735–6746

Sun H, Tao J, Hou M, Huang S, Chen S, Liang Z, Xie T, Wei Y, Xie X, Yoneyama K, Xu G, Zhang Y (2015) A strigolactone signal is required for adventitious root formation in rice. Ann Bot 115:1155–1162

Sun YK, Flematti GR, Smith SM, Waters MT (2016a) Reporter gene-facilitated detection of compounds in *Arabidopsis* leaf extracts that activate the karrikin signalling pathway. Front Plant Sci 7:1799

Sun H, Tao J, Gu P, Xu G, Zhang Y (2016b) The role of strigolactones in root development. Plant Signal Behav 11:e1110662

Sun H, Bi Y, Tao J, Huang S, Hou M, Xue R, Liang Z, Gu P, Yoneyama K, Xie X, Shen Q, Xu G, Zhang Y (2016c) Strigolactones are required for nitric oxide to induce root elongation in response to nitrogen and phosphate deficiencies in rice. Plant Cell Environ 39:1473–1484

Taniguchi M, Furutani M, Nishimura T, Nakamura M, Fushita T, Iijima K, Baba K, Tanaka H, Toyota M, Tasaka M, Morita MT (2017) The *Arabidopsis* LAZY1 family plays a key role in gravity signalling within statocytes and in branch angle control of roots and shoots. Plant Cell 29:1984–1999

Tilman D, Cassman KG, Matson PA, Naylor R, Polasky S (2002) Agricultural sustainability and intensive production practices. Nature 418:671–677

Torres-Vera R, Garcia JM, Pozo MJ, López-Ráez JA (2013) Do strigolactones contribute to plant defence? Mol Plant Pathol 15:211–216

Ueda H, Kusaba M (2015) Strigolactone regulates leaf senescence in concert with ethylene in *Arabidopsis*. Plant Physiol 169:138–147

Umehara M, Hanada A, Yoshida S, Akiyama K, Arite T, Takeda-Kamiya N, Magome H, Kamiya Y, Shirasu K, Yoneyama K, Kyozuka J, Yamaguchi S (2008) Inhibition of shoot branching by new terpenoid plant hormones. Nature 455:195–200

Umehara M, Hanada A, Magome H, Takeda-Kamiya N, Yamaguchi S (2010) Contribution of strigolactones to the inhibition of tiller bud outgrowth under phosphate deficiency in rice. Plant Cell Physiol 51:1118–1126

Visentin I, Vitali M, Ferrero M, Zhang Y, Ruyter-Spira C, Novák O, Strnad M, Lovisolo C, Schubert A, Cardinale F (2016) Low levels of strigolactones in roots as a component of the systemic signal of drought stress in tomato. New Phytol 212:954–963

Vogel JT, Walter MH, Giavalisco P, Lytovchenko A, Kohlen W, Charnikhova T, Simkin AJ, Goulet C, Strack D, Bouwmeester HJ, Fernie AR, Klee HJ (2010) SlCCD7 controls strigolactone biosynthesis, shoot branching and mycorrhiza-induced apocarotenoid formation in tomato. Plant J 61:300–311

Waldie T, McCulloch H, Leyser O (2014) Strigolactones and the control of plant development: lessons from shoot branching. Plant J 79:607–622

Walker CH, Bennett T (2017) Reassessing the evolution of strigolactone synthesis and signalling. BioRxiv:228320. https://doi.org/10.1101/228320

Walton A, Stes E, Goeminne G, Braem L, Vuylsteke M, Matthys C, De Cuyper C, Staes A, Vandenbussche J, Boyer FD, Vanholme R, Fromentin J, Boerjan W, Gevaert K, Goormachtig S (2016) The response of the root proteome to the synthetic strigolactone GR24 in *Arabidopsis*. Mol Cell Proteomics 15:2744–2755

Wallner ES, López-Salmerón V, Greb T (2016) Strigolactone versus gibberellin signalling: reemerging concepts? Planta 243:1339–1350

Wang L, Wang B, Jiang L, Liu X, Li X, Lu Z, Meng X, Wang Y, Smith SM, Li J (2015) Strigolactone signaling in *Arabidopsis* regulates shoot development by targeting D53-like SMXL repressor proteins for ubiquitination and degradation. Plant Cell 27:3128–3142

Wang Y, Sun S, Zhu W, Jia K, Yang H, Wang X (2013) Strigolactone/MAX2-induced degradation of brassinosteroid transcriptional effector BES1 regulates shoot branching. Dev Cell 27:681–688

Wang Y, Li J (2011) Branching in rice. Curr Opin Plant Biol 14:94–99

Waters MT, Gutjahr C, Bennett T, Nelson D (2017) Strigolactone signalling and evolution. Annu Rev Plant Biol 68:291–322

Waters MT, Nelson DC, Scaffidi A, Flematti GR, Sun YK, Dixon KW, Smith SM (2012) Specialisation within the DWARF14 protein family confers distinct responses to karrikins and strigolactones in *Arabidopsis*. Development 139:1285–1295

Yamada Y, Furusawa S, Nagasaka S, Shimomura K, Yamaguchi S, Umehara M (2014) Strigolactone signalling regulates rice leaf senescence in response to a phosphate deficiency. Planta 240:399–408

Yoneyama K, Xie X, Kim HI, Kisugi T, Nomura T, Sekimoto H, Yokota T (2012) How do nitrogen and phosphorus deficiencies affect strigolactone production and exudation? Planta 235:1197–1207

Young NF, Ferguson BJ, Antoniadi I, Bennett MH, Beveridge CA, Turnbull CG (2014) Conditional auxin response and differential cytokinin profiles in shoot branching mutants. Plant Physiol 165:1723–1736

Xu J, Zha M, Li Y, Ding Y, Chen L, Ding C, Wang S (2015) The interaction between nitrogen availability and auxin, cytokinin, and strigolactone in the control of shoot branching in rice (*Oryza sativa* L.). Plant Cell Rep 34:1647–1662

Zhang Y, Lv S, Wang G (2018) Strigolactones are common regulators in induction of stomatal closure in planta. Plant Signal Behav 23:e1444322

Zhou F, Lin Q, Zhu L, Ren Y, Zhou K, Shabek N, Wu F, Mao H, Dong W, Gan L, Ma W, Gao H, Chen J, Yang C, Wang D, Tan J, Zhang X, Guo X, Wang J, Jiang L, Liu X, Chen W, Chu J,

Yan C, Ueno K, Ito S, Asami T, Cheng Z, Wang J, Lei C, Zhai H, Wu C, Wang H, Zheng N, Wan J (2013) D14-SCF(D3)-dependent degradation of D53 regulates strigolactone signalling. Nature 504:406–410

Zou J, Zhang S, Zhang W, Li G, Chen Z, Zhai W, Zhao X, Pan X, Xie Q, Zhu L (2006) The rice *HIGH-TILLERING DWARF1* encoding an ortholog of *Arabidopsis* MAX3 is required for negative regulation of the outgrowth of axillary buds. Plant J 48:687–698

Chapter 3
Strigolactones and Parasitic Plants

Maurizio Vurro, Angela Boari, Benjamin Thiombiano, and Harro Bouwmeester

Abstract A parasitic plant is a flowering plant that attaches itself morphologically and physiologically to a host (another plant) by a modified root (the haustorium). Only about 25 out of the 270 genera of parasitic plants have a negative impact in agriculture and forestry and thus can be considered weeds. Among them, the most damaging root parasitic weeds belong to the genera *Orobanche* and *Phelipanche* (commonly named broomrapes) and *Striga* (witchweeds) (all belonging to the Orobanchaceae family). Considering the aims of the book, this chapter will focus only on this group of parasitic weeds, as in these plants strigolactones have a key role both in their life cycle, and in management strategies to control them. Distribution, agricultural importance and life cycle of these parasitic weeds are briefly introduced, after which we focus on the role of strigolactones in seed germination, parasite development, host specificity, plant nutrition and microbiome composition. Furthermore, some weed control approaches involving strigolactones are discussed.

Keywords Parasitic weeds · Orobanche · Phelipanche · Striga · Germination stimulants

3.1 Parasitic Plants

A parasitic plant is an angiosperm (flowering plant) that attaches itself morphologically and physiologically to its host (another plant) by a modified root (the haustorium). Depending on the host organ it is attached to, two main types of parasitic plants can be distinguished: stem parasites and root parasites. Stem parasites occur in several families and include mistletoes (*Viscum* spp.) and dodders (*Cuscuta* spp.),

M. Vurro (✉) · A. Boari
Institute of Sciences of Food Production, National Research Council, Bari, Italy
e-mail: maurizio.vurro@ispa.cnr.it

B. Thiombiano · H. Bouwmeester
Swammerdam Institute for Life Sciences, University of Amsterdam, Amsterdam, The Netherlands

© Springer Nature Switzerland AG 2019
H. Koltai, C. Prandi (eds.), *Strigolactones - Biology and Applications*,
https://doi.org/10.1007/978-3-030-12153-2_3

whereas root parasites are more common and belong to diverse taxonomic groups. Considering the aims of the book, attention will be given only to root parasites, and in particular to the root parasitic weeds of the Orobanchaceae, as strigolactones play a key role in their life cycle and in control strategies.

Parasitic plants can also be differentiated into obligate and facultative parasites. The former depend completely on their host, while the latter are capable of completing their life cycle without host contact and only attach to a suitable host if it is available. A further distinction can be made between holoparasites, which lack chlorophyll (and thus are non-photosynthetic), derive all their nutrition from their host and, therefore, are completely dependent upon the host to complete their life cycle; and hemiparasites, which contain chlorophyll (and hence are photosynthetic) and thus absolutely need the connection with the host only during part of their life cycle. All holoparasites are also obligate parasites. Although these definitions imply absolute categories, some parasitic plants display an intermediate behaviour between hemi- and holoparasitism, e.g. *Cuscuta* (dodder).

Many of the photosynthetic root hemiparasites are green with fully formed leaves, such as *Striga* spp. As the degree of parasitic dependence increases (i.e. the evolution from hemiparasitism to holoparasitism), profound changes occur in the morphology of the parasitic plant. In general, holoparasites tend to have leaves reduced to scales, succulent stems and primary (derived from the seedling radicle) and lateral (from developed roots) haustoria, whereas facultative parasites tend to have normal leaves and stems, and only lateral haustoria.

Only about 25 out of the 270 genera of parasitic plants have a negative impact on agriculture and forestry and thus can be considered weeds. Among them, the most damaging root parasitic weeds belong to the genera *Orobanche* and *Phelipanche* (commonly named broomrapes), *Striga* (witchweeds) and, to a lesser extent, *Alectra* and *Rhamphicarpa* (all of them belonging to the Orobanchaceae family). Among the weedy stem parasites, the most important genera are *Cuscuta* (dodder) of the Convolvulaceae family and *Arceuthobium* (dwarf mistletoe), *Viscum* and *Phoradendron* spp. (leafy mistletoes) of the Santalaceae family.

3.1.1   Importance

Witchweeds and broomrapes are responsible for enormous losses in major crops. Seven broomrape species are considered serious weeds, mainly in Europe, North Africa and Asia: *Phelipanche ramosa* (L.) Pomel, *Phelipanche aegyptiaca* (Pers.) Pomel, *Orobanche crenata* Forsk., *Orobanche cumana* Wallr., *Orobanche foetida* Poir., *Orobanche cernua* Loefl. and *Orobanche minor* Sm. With regard to the witchweeds, four species are considered very important weeds, present almost exclusively in Africa: *Striga hermonthica* (Del.) Benth., *Striga asiatica* (L.) Kuntze, *Striga aspera* Willd. and *Striga gesnerioides* (Willd.) Vatke (Parker 2013). Despite the large number of studies on their distribution and impact, in some countries losses and presence of parasitic weeds are probably underestimated because of the lack of data on minor crops or because farmers simply prefer to abandon risky crops in contaminated areas. Below follow brief descriptions of host range, distribution and severity of the main Orobanchaceae root parasitic weed species (see also Box 3.1).

Box 3.1 Some Key Features of the Most Troublesome Parasitic Weeds

	Flower colour	Stem height (cm)	Host range	Distribution
Orobanche crenata Forsk.	Generally whitish with purple veins	Up to 100	Wide. Many species mainly in Fabaceae and Apiaceae. Some Cucurbitaceae, Solanaceae, Lamiaceae, Ranunculaceae and Asteraceae	Predominantly around the Mediterranean including North Africa and into the Near East and Western Asia, with quite recent introductions into Sudan and Ethiopia
Orobanche cumana Wallr.	From white to pale blue	40–65	Specific to sunflower	SE Europe, Middle East and SW Asia. It is also present in China
Orobanche foetida Poir.	Dark red, yellowish or white at the base, shining dark red inside	20–70	Restricted to Fabaceae, wide within that family, mostly wild species but also faba bean, chickpea and vetch	Western Mediterranean: Morocco, Tunisia, Algeria and Libya in North Africa; Spain, Portugal and the Balearic Islands to the north. The weedy populations occur in Tunisia and Morocco
Orobanche cernua Loefl.	Whitish/pale yellow at the base, with deep blue/purple lips	Up to 35	Solanaceous crops, especially tomato, eggplant and tobacco, and, less commonly, potato	Southern Europe, Middle East, South Asia and Northern Africa, with possibly introduced infestations further south in Africa, in Niger, Sudan, Ethiopia, Kenya and Tanzania. On sandy beaches of South Australia

(continued)

Box 3.1 (continued)

	Flower colour	Stem height (cm)	Host range	Distribution
Orobanche minor Sm.	Mainly pale, whitish, with varying amounts of purple in the veins	Up to 50. In Ethiopia may exceed 100	Very wide. Many Fabaceae species (e.g. *Trifolium, Medicago, Arachis* spp.), Asteraceae (*Lactuca, Carthamus* spp.) and Apiaceae (*Daucus, Apium* spp.), Solanaceae and other families. Usually herbaceous but even woody hosts	Widely distributed. Native throughout most of Europe, other than the far north, Western Asia and Northern Africa, as far south as Ethiopia and Somalia. Sporadically introduced to Japan, New Zealand, Australia and several countries in North and South America
Phelipanche ramosa (L.) Pomel	From white at the base to pale blue or mauve to blue/purple on the lobes	Usually 10–30, occasionally 50	Many Solanaceae crops, especially tomato, eggplant and tobacco but also pepper and potato, and also Brassicaceae (rapeseed), Cannabaceae (hemp), Fabaceae (e.g. chickpea, clovers, groundnut, faba bean), Apiaceae (carrot, celery) and Asteraceae (lettuce, sunflower and ornamental species). Wild hosts in many families. Reported on onion but not on other monocots	Native distribution: Europe, Middle East, West Asia and North Africa south to Ethiopia and Somalia. New infestations recorded, e.g. Australia

(continued)

Box 3.1 (continued)

	Flower colour	Stem height (cm)	Host range	Distribution
Phelipanche aegyptiaca (Pers.) Pomel	See *P. ramosa*		The same host range as *P. ramosa*, in particular Solanaceae, Fabaceae, Apiaceae and Asteraceae. Wider range of Brassicaceae species and more important on Cucurbitaceae than *P. ramosa*. Occasionally on woody species	Distribution overlapping *P. ramosa* in South Europe, the Mediterranean and North Africa. Much further extended eastwards into South Asia and China
Striga hermonthica (Del.) Benth.	Pink (very occasionally white)	Up to 100, especially in Eastern Africa; about 50 in Western Africa	Most of the major tropical and subtropical cereals, especially sorghum, *Pennisetum*, millet and maize, but also upland rice, sugar cane and finger millet (*Eleusine coracana*)	Mainly northern sub-Saharan Africa from Senegal and Gambia in the west and to Sudan, Ethiopia and Kenya in the east. Except the Arabian Peninsula restricted to Africa
Striga asiatica (L.) Kuntze	Scarlet, occasionally yellow (or brick red in Ethiopia). White-flowered forms attack crops in South Asia	Usually 15–30	Host range as *S. hermonthica*, most notably maize and sorghum	Markedly differing from *S. hermonthica*, being predominantly in Eastern and Southern Africa. The two species overlap in Kenya and Tanzania but rarely occur together

(continued)

Box 3.1 (continued)

	Flower colour	Stem height (cm)	Host range	Distribution
Striga aspera Willd.	Resembling *S. hermonthica* in general appearance and flower colour, usually somewhat smaller		Most of the warm-climate cereals. Less common on sorghum and pearl millet and more common on rice and sugar cane than *S. hermonthica*	Mainly in West Africa but also eastwards to Sudan and south to Malawi
Striga gesnerioides (Willd.) Vatke	From white to mauve to purple	12–30	Only dicotyledons; cowpea is the main host	Mainly in Africa (West in particular) but also South and SE Asia
Alectra vogelii Benth	Yellow, sometimes with purple streaks	30–45	Various Fabaceous crops: cowpea is the main, groundnut, soybean and other legumes	A number of West African countries (especially Nigeria and Burkina Faso) and in other countries of Southern and Eastern Africa

3.1.2   Main Orobanchaceae Root Parasitic Weeds

***Phelipanche ramosa* (L.) Pomel.**   The host range of *P. ramosa* is extremely wide. It preferably parasitizes Solanaceae species (tomato, potato and tobacco in particular), Asteraceae (e.g. lettuce and sunflower), Brassicaceae (cabbage, rapeseed), Cannabaceae (hemp), Fabaceae (e.g. chickpea and faba bean) and Apiaceae (carrot, celery). Rapeseed, cabbage and hemp are now increasingly affected (Parker 2013). *P. ramosa* is native around the Mediterranean basin and originally infested crops only in Europe, the Middle East, West Asia and North Africa, but new infestations have been reported, e.g. in Australia (Warren 2006). The most severe yield losses in tomato vary between 30 and 50% in Slovakia (Cagáň and Tóth 2003) to over 80% in Chile (Díaz et al. 2006). In Sudan, heavy infestations caused the closure of tomato juicing factories (Babiker et al. 2007). Other countries in which tomato and/or eggplant have been seriously affected include, among others, Italy, Greece, Iran, Hungary and Cuba (Parker 2013). *P. ramosa* also attacks tobacco in Moldova (Timus and Croitoru 2007), Cuba and Italy (Zonno et al. 2000), and rapeseed in France (Gibot-Leclerc 2003).

Phelipanche aegyptiaca (Pers.) **Pomel.** *Phelipanche aegyptiaca* has a host range similar to that of *P. ramosa*, attacking in particular the Solanaceae tomato, potato, eggplant and tobacco, and crops in the Fabaceae, Apiaceae and Asteraceae, too. It seems to have a wider range of Brassicaceae and to be more important on Cucurbitaceae than *P. ramosa*, being also occasionally occurring on woody species (Eizenberg et al. 2002). Compared to *P. ramosa*, it has almost the same geographical distribution in the Mediterranean countries, South Europe and Northern Africa, but it extends much further eastwards into South Asia and China. The effects of *P. aegyptiaca* on the host are the same of those caused by *P. ramosa*. Damage can be very severe, as advised on lentil in Turkey (Bülbül et al. 2009) or on *Eruca sativa* in India (Bedi et al. 1997), amounting to around 40%. In Iran over 70% yield loss was reported in potato (Motazedi et al. 2010) and severe losses in water melon (Parker and Riches 1993).

Orobanche crenata **Forsk.** *Orobanche crenata* has a moderately wide host range including species in the Fabaceae and Apiaceae but also some in the Cucurbitaceae, Solanaceae, Lamiaceae, Ranunculaceae and Asteraceae (Musselman and Parker 1982). Its native distribution is predominantly around the Mediterranean Sea including North Africa and into the Near East and Western Asia. *O. crenata* is especially important all around the Mediterranean Sea where it infests the most important legume crops, particularly faba bean (Fig. 3.1), lentil and chickpea but also carrot. Around 180,000 ha were estimated to be infested in Morocco, Portugal, Spain and Syria, representing 50–70% of the areas of these crops grown in those countries. Yield losses amounting to 33% in Egypt, from 50 to 100% in Malta, and up to 70% in Turkey were estimated to occur (Sauerborn 1991).

Orobanche cumana **Wallr.** *Orobanche cumana* is one of the most important biological constraints of sunflower production and is particularly important in Russia, Ukraine, Moldavia, Romania, Turkey, Bulgaria, Spain, Israel and Hungary but occurs also in Syria and Egypt and along the North African coast (Parker 1994). Areas of sunflower affected have been estimated at 40,000 ha in Greece and 20,000 ha in China, with around 60% and 20–50% losses, respectively. In Turkey, over 50% of the crop area was moderately infested in spite of the use of resistant varieties. Earlier studies estimated a reduction by 37% of the area where sunflower was grown because of heavy infestation in the former Yugoslavia before the introduction of resistant varieties (Sauerborn 1991). *O. cumana* in sunflower has been the subject of extensive research for the breeding of resistant varieties, which has provided only a temporary alleviation of the problem, as this resulted in the development of more virulent races shortly after the introduction of the resistant varieties.

Orobanche foetida **Poir.** The native range of *O. foetida* is limited to the Western Mediterranean countries, e.g. Morocco and Tunisia in North Africa and Spain and Portugal in Europe. Although *O. foetida* occurs on a number of wild leguminous hosts, it is only a significant problem in faba bean, chickpea and vetch, for example, in Tunisia, since the last couple of decades (Román et al. 2007). This should be

Fig. 3.1 *Orobanche crenata* plant attacking faba ben

particularly worrying in the other countries around the Western Mediterranean, where it still occurs only on wild hosts. Damage to faba bean can be severe, resulting in losses of over 90% of seed yield (Abbes et al. 2007).

***Orobanche cernua* Loefl.** *Orobanche cernua* is almost exclusively a parasite of Solanaceae, especially tomato, tobacco and eggplant. Its distribution extends from Southern and Eastern Europe to North Africa but also from Asia to Australia. *O. cernua* is a very serious problem on tobacco in Asian countries, e.g. Pakistan, Iran and India. In the latter country, on thousands of hectares, severe infestation in tobacco has been reported causing large qualitative and quantitative yield losses. In tomato, severe infestations have been reported in countries such as Ethiopia, Israel and Kenya (Parker 1994).

***Orobanche minor* Sm.** *Orobanche minor* is a smaller problem compared to the other broomrapes. It is broadly distributed throughout most of Europe (except the Northern countries) and the Middle East and also along the western coast of North Africa. It has also been sporadically introduced to other countries, e.g. Japan or North- and South-American countries. Clover and alfalfa are the main crops affected, although not severely. Hosts are usually herbaceous but can also be

Fig. 3.2 *Striga hermonthica*

woody, e.g. pecan. Reductions in total host weight up to 50% were reported (Lins et al. 2006), with problems for the quality of the crop, that cannot be sold due to the contamination with seeds of the parasite (Mallory-Smith and Colquhoun 2012).

***Striga hermonthica* (Del.) Benth.** This is the most damaging of the *Striga* species (Fig. 3.2), occurring mainly in northern sub-Saharan African countries such as Senegal, Sudan, Ethiopia, Mali, Benin, Nigeria and Kenya. It occurs in the Arabian Peninsula but is otherwise restricted to Africa. *S. hermonthica* is considered the most serious worldwide parasitic weed, with an estimated affected area amounting to many millions of hectares (Sauerborn 1991; Parker 2009, 2013). Most of the major tropical and subtropical cereals are affected, in particular sorghum, millet and maize but also upland rice and sugar cane. *S. hermonthica* is a photosynthetic species, although not very efficient (Press et al. 1987). The effects of an infection are visible well before emergence and consist in stunting of the host shoot (Parker 1994) and chlorotic blotching of its foliage. The overall effect on the host can be devastating and lead to total crop failure. Losses of maize in Kenya may reach 80% in case of heavy infestation (Manyong et al. 2007). Estimates for all cereals in 1991 varied from 40 to 50% in Ghana, Cameroon and Nigeria to over 70% in Benin and Gambia (Sauerborn 1991; Gressel et al. 2004; Labrada 2007; Ejeta 2007; Scholes and Press

2008). In countries such as Togo, Mali and Nigeria, the infested area is estimated to be around 40%, reaching over 60% in Benin (De Groote et al. 2008), and even over 80% in north-east Nigeria (Dugje et al. 2006). Across the whole of Africa between 50 and 300 million ha are estimated to be infested by the parasite.

Striga asiatica **(L.) Kuntze.** *Striga asiatica* attacks the same crops as *S. hermonthica* and in particular maize and sorghum. It is distributed predominantly in Eastern and Southern Africa, with an overlap of the two species in Kenya and Tanzania. However, they rarely occur together. Although the damage caused by *S. asiatica* is similar to *S. hermonthica*, it represents a lower economic problem worldwide compared to the latter. The physiological effects on the host are stunting, a change in host root-to-shoot ratio, reduction of host photosynthesis and wilting even under moist conditions. Crop losses between 10 and 40% are common. Up to 80% losses were estimated to occur in maize in several Southern African countries (De Groote et al. 2008).

Striga aspera **Willd.** *Striga aspera* resembles *S. hermonthica* in the general appearance and the effects on the parasitized crops. It can attack most of the warm-climate cereals, but it is less common on sorghum and pearl millet and somewhat more common on rice and sugar cane than *S. hermonthica* (Parker and Riches 1993). *S. aspera* occurs mainly in West Africa but also more to the east in Sudan and to the south in Malawi. A reduction of around 50% in rice yield as a consequence of *S. aspera* infection has been recorded (Johnson et al. 1997).

Striga gesnerioides **(Willd.) Vatke.** This autogamous species has different races, differing from each other in host species and/or genotype range and to some extent in morphology (e.g. number of branches, colour of stem and corolla). Among the *Striga* species, it is the most widely distributed (Mohamed et al. 2001), being particularly important on cowpea in West Africa, where crop losses can exceed 50%. Sweet potato, tobacco and a number of other wild species can also be attacked by *S. gesnerioides* races.

Alectra vogelii **Benth.** As the related *Striga* spp., *A. vogelii* is an obligate hemiparasite having green foliage. Cowpea represents its main host, but a number of other legume crops, such as groundnut and soybean, can be attacked, too. This species occurs across much of Africa, with cowpea seriously affected in several West African countries, especially Nigeria and Burkina Faso. Damage can be very severe, and even complete yield losses have been reported (Emechebe et al. 1991).

Consideration. Parasitic weeds may represent an increasing problem in agriculture, due to changes in crop production and rotations, in response to global warming, and due to socioeconomic and political changes. For example, changes in the dietary wishes of consumers and more attention for the environment and the preservation of soil fertility are favouring the increase in legume production area in Western Europe, which in combination with a warmer climate could increase the risk of *O. crenata* establishment.

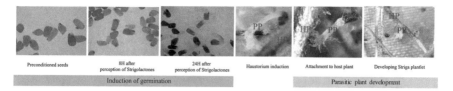

Fig. 3.3 *Striga* developmental cycle (*PP* parasitic plant, *HP* host plant)

The production of biofuels all over the world is dramatically increasing in this last decade. Some of the crops used for biofuel production, i.e. oilseed rape and sunflower, are broomrape hosts and have started to be grown in new areas, thus increasing the potential area of broomrape hosts and therefore the risk of an increase in the infested area.

Other problems could be represented both by the introduction of new crops, where traditional non-host crops potentially are replaced by host crops, and by the introduction of parasitic weeds in noninfested areas, due to global warming changes and international (sometimes not checked or tracked) trading and traffic.

3.1.3 Life Cycle

Although *Orobanche* and *Phelipanche* spp. are obligate holoparasites, whereas *Striga* spp. are obligate hemiparasites, the species of these three genera share many similarities. Their flower shoots have a spike, bearing from 10 to 20 flowers in most species, to even 100 or more. Fruits are capsules, each producing between around 500 extremely small (200–400 µm) seeds (Joel et al. 2007). Each plant can produce several tens of capsules and thus up to 1 million seeds. The life cycle of these parasites starts with seed germination, followed by the attachment to the host, which represents the beginning of the parasitic life phase (Fig. 3.3).

Some preparatory metabolic processes take place before the seed can react to stimuli and germinate. This preparatory phase, known as "conditioning", is a complex metabolic and developmental process that consists of a series of events, each crucial for achieving germination. When a ripe seed comes in contact with water, it imbibes in less than 1 day; however, a moist environment is required for several days together with a suitable temperature in order to make the imbibed seed ready to perceive a chemical stimulus to germinate (see next sections). If conditioned seeds are not exposed to a germination stimulant and germination does not occur, their sensitivity gradually decreases again, and the seeds enter into secondary dormancy (Matusova et al. 2004). Upon germination, the radicle emerges from the seed reaching a length of a few mm up to 1 cm long (Fig. 3.4). Upon contact with a host root, the radicle develops intrusive cells that penetrate the root (Losner-Goshen et al. 1998) forming the haustorium, a physiological bridge between the vascular system of the host and that of the parasite. In *Striga* spp., the haustorium establishes a

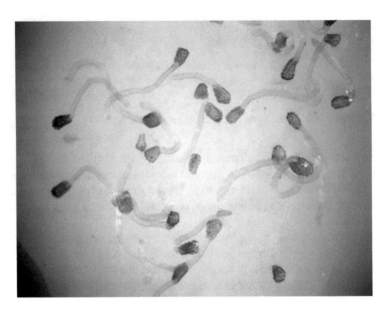

Fig. 3.4 Germinated seeds of *Phelipanche ramosa*

xylem–xylem connection with the host from where it can withdraw water and nutrients. *Phelipanche* and *Orobanche* spp. form connections with both phloem and xylem (Westwood 2013). If the germinating seed fails to reach a host, it will die. The haustorium first serves as an attachment organ and structure to penetrate the host tissues and then becomes an organ that absorbs water and nutrients from the host, the real beginning of the parasitic phase. Therefore, this phase is essential and crucial to any further development of the parasite. After the establishment of the haustorium, the parasite develops a tubercle, which is the juvenile parasite that accumulates water and nutrients. Subsequently, the parasite develops a shoot that emerges from the soil, produces flowers and set seeds that can remain vital over decades in the soil, thus completing its life cycle.

The production of many tiny seeds increases the dispersion of the parasite into the soil profile, and therefore the chance to meet the roots of a suitable host that will induce germination and allow attachment. Host plant density and root shape can result in improved reproduction conditions for the parasites, increasing the probability of infecting the crop. This supports the build-up of enormous seed banks, which represent one of the main problems in parasitic weed management, as the seeds may remain dormant in the soil for many years, also if a host is not grown.

In *Orobanche* and *Phelipanche* spp. the reduction in biomass of infected hosts can be largely explained by the biomass accumulation of the parasite. However, the strong depression of the host growth caused by *Striga* spp. is only partially correlated with the increase in parasite biomass. The negative impact on host growth in *Striga*-infected plants can already be observed even before the parasite has emerged from the soil, suggesting that *Striga* spp. have a pathological or phytotoxic effect on the host plant.

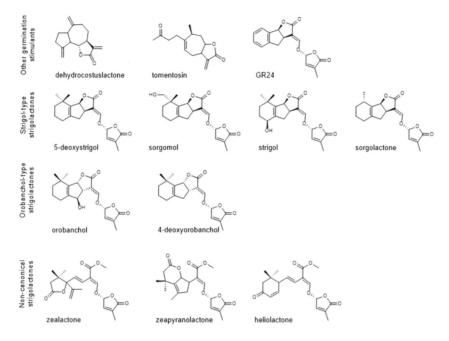

Fig. 3.5 Chemical structures of some of the SLs mentioned in the chapter

3.2 Role of SLs in Seed Germination and Parasite Development

3.2.1 SLs and Seed Germination

Several different compound classes have been described acting as germination stimulants in many different roots parasitic plant species. Examples are isothiocyanate, which stimulates the germination of *P. ramosa* that infects rapeseed, and dehydrocostus lactone and tomentosin (Fig. 3.5) which stimulate the germination of *O. cumana* that infects sunflower (Pérez-de-Luque et al. 2000; Auger et al. 2012). Strigolactones (SLs)—which are biosynthetically derived from the carotenoids (Matusova et al. 2005)—are, however, the major class of germination stimulants and have been shown to induce the germination of many of the Orobanchaceae root parasitic plants (Bouwmeester et al. 2003; Yoneyama et al. 2010). SLs are actively transported into the rhizosphere by a range of plant species and were—decades after their discovery as germination stimulants—shown to play an important role also in the interaction of plants with arbuscular mycorrhizal (AM) fungi. On top of that, they also have an endogenous signalling role in plants and are a new class of plant hormones controlling shoot branching and root development (Domagalska and Leyser 2011; Koltai 2011; Ruyter-Spira et al. 2013). Root exudate analysis of parasitic plants hosts shows that they may contain different classes of SLs (Wang and Bouwmeester 2018). Whereas in exudates of sorghum, the main SLs are 5-deoxystrigol, strigol, sorgomol and sorgolactone (all strigol-type strigolactones) (Fig. 3.5), those in tomato are orobanchol-type strigolactones such as orobanchol, solanacol and didehydro-orobanchol isomers (Fig. 3.5) (Wang and Bouwmeester 2018). In addition to these canonical SLs, species such as maize and sunflower also produce so-called noncanonical SLs, such as zealactone, zeapyranolactone and heliolactone (Fig. 3.5) (Ueno et al. 2014; Charnikhova et al. 2017, 2018) (Also see Chap. 1). In some species and/or genotypes, these different categories also occur together, such as in certain sorghum genotypes that produce orobanchol as well as 5-deoxystrigol (Gobena et al. 2017).

These exuded SLs are essential signalling molecules in the parasitic plant life cycle as their detection by responsive (conditioned; see above) parasitic plant seeds results in the induction of germination (Fig. 3.4). In general, the configuration of SLs is determining their germination stimulatory activity towards the different species of parasitic plants. For example, seeds of *S. gesnerioides* are more sensitive to orobanchol-type SLs, while *S. hermonthica* generally is more responsive to strigol-type SLs (Ueno et al. 2011a, b; Gobena et al. 2017).

3.2.2 SLs and Parasitic Plant Development

As described above, after germination of the parasite the infection process of the host plant continues with the formation of the haustorium, induced by haustorium inducing factors released by the host root (Riopel and Timko 1995), which results in a connection between the parasitic plant and the host plant. At this stage the host plant becomes a source of nutrients for the parasitic plant, as well as the exchange of signalling molecules between the two (Press et al. 1987; Těšitel et al. 2010; Liu et al. 2014; Lei 2017; Spallek et al. 2017).

Plant hormones such as auxin have been suggested to play a role in the successful establishment of the connection between parasite and host (Bar-Nun et al. 2008) and defence hormones such as salicylic acid and jasmonic acid have been implicated as possible defence inducers (Letousey et al. 2007; Dita et al. 2009; Torres-Vera et al. 2014). As SLs are also a plant hormone, a possible role of SLs from the host on parasitic plant development would not be unlikely. Indeed, a number of studies have shown the importance of host plant SLs, also after germination, in the infection process. Silencing of CCD8, one of the core SL biosynthesis pathway genes, in tomato resulted in a stronger infection by *P. ramosa* upon infection with pre-germinated seeds (Cheng et al. 2017). The authors proposed that this may be caused by a modification in the auxin levels as a result of the lower SL production in the mutant, which would facilitate the formation of a vascular connection with the host or by a reduction in the levels of defence-related hormones such as salicylic acid and jasmonic acid. Also in rice, it seems that lower SL production results in decreased induction of germination but in increased impact of the infection (lower tolerance) after attachment (Cardoso et al. 2014).

Although SLs—with the exception of host SLs (Liu et al. 2014)—have not been detected in any of the root parasitic plant species so far, there are strong indications that they can produce them as they have and express all the SL biosynthetic genes (Liu et al. 2014; Das et al. 2015). Clear evidence of the involvement of endogenous SLs of parasitic plants in the infection comes from the work of Aly et al. They showed that trans-silencing CCD7 and CCD8 genes using VIGS in *P. ramosa* resulted in a strong reduction in the formation of tubercles (by more than 90%) during the infection process (Aly et al. 2014). In addition to their own SLs, there is evidence that host SLs are transported from the host to *Striga* (Liu et al. 2014). It is unknown whether these also have an effect on the development of the parasite.

3.3 SLs and Host Specificity

Among parasitic plants, a certain degree of host specificity can be observed (see Sect. 3.1.2). For some parasitic plant species, the host range is very narrow, such as *O. cumana* on sunflower. For others the host range is very wide. For example, *P. ramosa* can infect Solanaceae including tomato and potato and Brassicaceae including cabbage and oilseed rape (Gibot-Leclerc et al. 2016; Perronne et al. 2017). This broad host range does, however, seem to coincide with host specificity in ecotypes of one species. For example, even if they are able to colonize different hosts, the exposure of *P. ramosa* seeds to exudates from different host species led to different germination rates (Perronne et al. 2017). The same holds for *S. hermonthica*. This species can infect a large variety of cereals (maize, sorghum, millet, rice), but there are ecotypes of the species which are more successful on millet than on sorghum and maize and vice versa (Kim et al. 1994; Mohemed et al. 2018). The question whether this host specificity (including in ecotypes) is due to germination stimulants is intriguing. In sunflower this seems to be the case, as *O. cumana* preferentially germinates with dehydrocostus lactone (Fig. 3.5), a molecule present in the exudate of sunflower and not in response to SLs (Auger et al. 2012). For *S. hermonthica*, SLs seem to be the major germination stimulant, and there are indications that SL composition plays a role in host specificity. The *S. hermonthica* sorghum ecotype germinates much less well with a millet exudate and vice versa (Mohemed et al. 2018). Work on sorghum, maize and rice aiming at the identification of varieties resistant to *S. hermonthica* points to a higher susceptibility for cultivars producing more 5-deoxystrigol (Jamil et al. 2011a; Yoneyama et al. 2015; Mohemed et al. 2018). Conversely, sorghum genotypes that produce more orobanchol than 5-deoxystrigol are much less sensitive to *S. hermonthica* (Gobena et al. 2017; Mohemed et al. 2018).

The individual evaluation of SLs for their ability to induce parasitic plant seed germination has confirmed that seeds of different species differentially respond to different SLs (Wang and Bouwmeester 2018). For example, *O. minor* germination can be achieved with about 200 times less *ent-2′-epi*-orobanchol when compared to *S. hermonthica* (Ueno et al. 2011b). In a similar way, exposure of *S. gesnerioides* and *S. hermonthica* to the same concentration of 5-deoxystrigol induced only germination of the latter (Ueno et al. 2011a).

3.4   Role of SLs in Belowground Interactions of the Host

3.4.1   Plant Nutrition (Phosphate, Nitrogen)

When plants are subjected to stress such as phosphate or nitrogen deficiency, they use several adaptation strategies, of which the most important are the modification of the root and shoot architecture, the establishment of favourable interactions with microorganisms and the modification of the rhizosphere pH (Bouwmeester et al. 2007; Péret et al. 2011; Yoneyama et al. 2012; Kumar et al. 2015). All these mechanisms aim to increase the proportion of nutrients available for the plants. Interestingly, when plants are grown on nitrate, but especially phosphate, deficient media, an increase in the production of SLs is induced (López-Ráez et al. 2008; Yoneyama et al. 2012, 2015; Marzec et al. 2013; Ito et al. 2016). There are several indications that this upregulation of SL production plays a role in the adaptation of plants to the low nutrient conditions. In the absence of phosphate, for example, plants favour the production of lateral roots (Péret et al. 2011) in order to increase the surface in contact with the soil. Auxin has been shown to play an important role in this adaptation as it is implicated in the initiation of lateral root primordia and the emergence of lateral roots (Chiou and Lin 2011; Sun et al. 2014). In addition to auxin, more and more work is also pointing to a role for SLs in the adaptation of root architecture to phosphate deficiency (Ruyter-Spira et al. 2011; Sun et al. 2014; Kumar et al. 2015). Under phosphate deficiency, a cross talk between SLs and auxin is taking place which results in an increase in lateral root density (Ruyter-Spira et al. 2011). This change in root architecture may also have an effect on parasitic plant infection, as it seems to increase the chance of a host root to come into the vicinity of seeds of the parasite. The increased production and exudation of SLs under these conditions also trigger the improved colonization of the roots by symbiotic microorganisms (see Sect. 3.4.2) but also results in increased germination of parasitic plant seeds and therefore in higher infection (Jamil et al. 2012, 2013, 2014a, b).

3.4.2   Microorganisms

SLs are also actors in the structuring of the biotic environment around the roots of plants. They promote the effectiveness of colonization by arbuscular mycorrhizal fungi (AM fungi), as hyphal branching factors (Akiyama et al. 2005; Besserer et al. 2006). In addition to the symbiotic interaction with AM fungi, SLs have also been shown to play a role in nodulation. A pea *rms1* mutant showing undetectable SL levels in roots tissue and in root exudates displayed a strongly reduced nodule number that was 40% lower than in the *wt* (Foo et al. 2013). In soybean a decrease in nodulation was observed in *GmMAX3b* knockdown lines, while overexpression of the same gene in transgenic hairy roots displayed an increased nodule number (Haq et al. 2017). It is yet unclear whether this is due to a signalling function of the SLs or their hormonal effect. SLs are not the only chemicals that are exuded by plants. The rhizosphere is a zone surrounding the plant roots, which has a very large chemical diversity. The exuded molecules serve not only as a carbon source for microorganisms but also play a role as signalling molecules. This chemical diversity is likely the engine of recruitment and selection of specific microorganisms. One of the most studied cases today remains that of phenylpropanoids that are involved in both symbiotic (Abdel-Lateif et al. 2012; Liu and Murray 2016) and allelopathic mechanisms (Bais et al. 2006).

Do SLs also play a role in microbiome recruitment? Recent work on sorghum demonstrates the ability of different genotypes to recruit different bacterial communities from the soil in which they are grown. The *Striga*-resistant genotype SRN39 has a different SL profile as other sorghum genotypes (Gobena et al. 2017) and recruited a microbiome that was different from that of the others (Schlemper et al. 2017). An intriguing question is if these changes at the microbiome level have an effect on the infection of the host by parasitic plants. Indeed, from a Kenyan *Striga*, suppressive soil bacteria could be isolated that induced up to 45% of decay in *Striga* seeds (Neondo et al. 2017). Other mechanisms by which soil microorganisms could suppress parasitic plants include the production of germination inhibiting factors, inhibitors of radicle growth and haustorium formation, strengthening the vigour of the host plant by activating plant defence mechanisms or competitive utilization of signalling molecules inducing parasitic plant seed germination.

3.5   SLs and Parasitic Weed Management

The main difficulties in controlling parasitic Orobanchaceae weeds are on the one hand related to the intrinsic characteristics of the parasitism (i.e. the physic and physiological connection between host and parasite) and on the other hand to the properties of their seeds (i.e. the enormous number produced by each plant, the minute size, their longevity and the easy dispersal). The first characteristics hamper all the classical interventions attempting to control the weed without damaging the host (e.g. mechanical, physical and chemical). This causes a rapid increase in the soil seed bank, even when the original infested area is very limited, or even when only a few plants are left after effective management practices. Containment of infested areas and prevention of seed distribution should therefore be a major objective of parasitic weed management strategies, in addition to direct control interventions against the parasites (Rubiales et al. 2009). In this chapter, we will not review all possible control and management strategies of parasitic weeds but focus on methods—which are already used or can potentially be developed—that are based on the importance of SLs in the life cycle of these parasites. Indeed, there are several strategies of weed management focussing on the SLs, trying to avoid the stimulation of germination, or conversely to favour it, in the absence of a host. These practices are briefly considered in the next sections.

3.5.1   Trap and Catch Crops

The aim of the use of trap and catch crops is not to directly control the parasitic weeds, but rather to reduce the infestation over time, by reducing the seed bank in the soil. Trap crops are non(false)-host crops of which the roots release strigolactones, thus stimulating parasitic plant seed germination, but—since they are not a host—without allowing further development of the parasite, by impeding a viable connection of the haustorium to the host root (Parker and Riches 1993). This effect is also defined as "suicidal" germination. Trap crops can be used both for intercropping, i.e. by growing it in between the main crop, and as a main crop on itself. Besides its main effect, the induction of seed germination, a non-host crop can potentially also contribute to parasitic weed control by providing shade and reducing soil temperature (as a cover crop).

One of the best examples of an effective intercrop species with proven success in *S. hermonthica* suppression is *Desmodium uncinatum* Jacq. (Pickett et al. 2010; Hooper et al. 2010). This forage legume not only improves the soil fertility but also causes suicidal seed germination and inhibition of the parasite attachments to the host roots, by producing simultaneously both stimulatory and inhibitory flavonoid compounds in their root exudates (Khan et al. 2010). *Striga* may also be controlled by rotating or intercropping the cereal crop with other plant species, e.g. groundnut (*Arachis hypogea*) (Carson 1989), pigeon pea (*Cajanus cajan*) (Oswald and Ransom 2001) or cotton (*Gossypium* spp.) (Swanton and Booth 2004).

Several trap crops have been reported to reduce broomrape seed banks (even if some of them were effective only under controlled conditions), such as, sorghum (*Sorghum bicolor*), flax (*Linum usitatissimum*) and soybean (*Glycine max*) (Al-Menoufi 1989; Saxena et al. 1994; Kleifeld et al. 1994; Abebe et al. 2005). Other examples of effective broomrape trap crops include flax against *O. crenata*; different wheat cultivars against *O. minor*, radish, linseed, fennel and cumin against *P. aegyptiaca*; and hybrid maize against *O. cumana* (Gbèhounou and Adango 2003; Acharya 2014; Aksoy et al. 2015).

Conversely, catch crops are host plants that also produce strigolactones but do allow attachment by the parasite. In this case, the crop is simply removed from the field after the parasite seeds have germinated (and possibly attached), but before flowering and seed dispersal of the parasite are initiated. Important crops reported as potential catch crops for broomrape control are faba bean (*Vicia faba*), field mustard (*Brassica campestris*), white mustard (*Sinapis alba*), lentil (*Lens culinaris*), berseem clover (*Trifolium alexandrinum*) and fenugreek (*Trigonella foenum-graecum*) (Sauerborn and Saxena 1986; Parker and Riches 1993; Kleifeld et al. 1994; Dhanapal et al. 1996; Acharya et al. 2002; Fernández-Aparicio et al. 2008, 2010).

3.5.2 Suicidal Germination by SLs, Analogues and Mimics

As an alternative to trap and catch crops, which require that they are grown for a certain period of time on the contaminated field, suicidal germination can potentially also be provoked by applying compounds with stimulatory activity directly to the field. The parasitic seeds would germinate in the absence of a host and would hence not survive. Generally, the most active molecules inducing seed germination are the naturally occurring SLs, including 5-deoxystrigol and orobanchol. Unfortunately, the structures of these natural SLs are rather complex. As a result, synthesis of these SLs for effective field applications is not feasible. Therefore, alternative approaches to produce germination stimulants have been explored. Examples are the synthesis of simpler and cheaper SL analogues, the use of more easily available, natural compounds from other sources and the use of other compounds from whatever origin with stimulatory activity.

The first encouraging attempts to achieve suicidal germination with synthetic SLs in the field were obtained by using GR7 (Babiker and Hamdoun 1982) (this is GR24 (Fig. 3.5) lacking the aromatic A-ring, see Chap. 6). Interesting results in field experiments were also reported using Nijmegen-1 as SL analogue in tobacco infested by *O. cumana* (Zwanenburg et al. 2009). Although they proved to work effectively in reducing the parasitic seed load and protecting the host plants subsequently grown in the affected field, problems regarding their production cost, potential off-target effects in the soil and low stability remain to be solved (Zwanenburg and Pospíšil 2013; Zwanenburg et al. 2016). Some SL analogues have been used with promising results in pot experiments (Kgosi et al. 2012) formulated in an emulsion, which prevented hydrolysis and leaching down to lower soil layers.

Natural products that have similar activity as SLs have been isolated from a variety of sources. For example, dihydrosorgoleone was identified in the root exudate of sorghum and was shown to have germination stimulating activity for *S. asiatica* (Chang et al. 1986); dehydrocostus lactone (Fig. 3.5) was identified in the root exudates of sunflower as the natural germination stimulant for *O. cumana*, a root parasite specific of sunflower (Joel et al. 2011). Peagol and peagoldione, which bear some structural similarities to the SLs, were isolated from pea (*Pisum sativum*) root exudates and exhibited germination stimulatory activity in particular on *O. foetida* (only peagol) and *P. aegyptiaca* (Evidente et al. 2009), whereas soyasapogenol B and trans-22-dehydrocampesterol were isolated from common vetch (*Vicia sativa*) exudates and stimulated germination of different broomrape species (Evidente et al. 2011). However, most of these compounds proved to have only a modest stimulatory activity only under lab conditions. Thus, their use for controlling parasitic weeds is very far from being put into practice.

In recent years, a group of compounds not having the SL bioactiphore has been described. These compounds are based on the D-ring with an appropriate substituent at C-5. These compounds are referred to as SL mimics. Currently, two types of SL mimics are available. The first has a substituted phenyloxy group at C-5. Para-bromo-phenyloxy butenolide is weakly active on *S. hermonthica*. This group of phenoxy-substituted butenolides are also called debranones (debranching furanones). Synthetically, these SL mimics are very easy to prepare from either bromo butenolide or hydroxy butenolide, opening up new possibilities for a practical use of these compounds for clearing of parasitic weed infested fields (Zwanenburg et al. 2016). The second group of compounds, which was reported almost at the same time, contains an aroyloxy group at C-5. These SL mimics are modestly active as germination agents for *S. hermonthica* seeds but are remarkably active for *O. cernua* seeds (Zwanenburg et al. 2016). A carbamate with moderate germination-inducing activity and facile preparation, named T-010, formulated as a 10% wettable powder, was evaluated for germination-inducing activity towards the purple witchweed (*S. hermonthica*) in greenhouse and field experiments showing very promising preliminary effects (Samejima et al. 2016).

A compound not related to SLs, used for control of *Striga* spp., is ethylene (Rodenburg et al. 2005). It is injected into the soil and provokes seed germination of *Striga* spp. and successive death due to the absence of a suitable host. Although ethylene application has been successfully employed as part of the *Striga* eradication programme in the USA (Tasker and Westwood 2012), the practice is very expensive—so not suitable for use in the developing world—and its use not a guarantee for total eradication.

3.5.3 SL Degradation

A different approach for controlling root parasitic weeds would be the degradation of the SLs soon after they are released into the soil by the host roots, and before the stimulatory signal reaches the seeds of the parasite. The ultimate goal of this approach would not be a reduction of the seed bank over time, but rather to enable growing susceptible crops on infested fields. To achieve this, both chemical and biological approaches were explored. For the chemical approach, borax was used, an inexpensive and eco-friendly salt. It was successfully demonstrated under laboratory conditions that borax can be used to decompose germination stimulants prior to their interaction with seeds of parasitic weeds (Kannan and Zwanenburg 2014). For practical field applications, formulation of borax would be necessary, and the method would require optimization because in the long run, its continued use could give rise to too high boron concentrations in the soil, resulting in undesirable soil intoxication. The possibility to prepare a film of borax emulsion, formulated with a salt, around the seeds of the parasites has been suggested (Kannan and Zwanenburg 2014). This would ensure that no active stimulant would reach the seeds even if some of the stimulants would escape decomposition after being exuded. This would be an example of double gatekeeping: decomposition of the stimulant when exuded from the roots and when approaching the seeds of the parasite. Another agent for rapid decomposition of SLs could be a renowned nucleophilic agent, namely thiourea, acting in a similar way as borax. It can be easily formulated, is an inexpensive eco-friendly compound, a bio-regulatory molecule for plant growth stimulation, and also acts as an antioxidant in plant protection (Kannan and Zwanenburg 2014).

SLs have been reported to be present in the root exudates of a wide range of different plant species (see above), and thus it would not be surprising if these compounds also act as signals for microorganisms other than AM fungi that could be beneficial to the host (e.g. ectomycorrhizal fungi, biocontrol agents, biofertilizers, resistance inducers) and phytopathogenic to the parasites (also see above). These aspects could be highly interesting from a practical point of view, allowing novel approaches for parasitic plant management. For example, the potential of some beneficial microorganisms to metabolize SLs and to be rhizosphere competent (i.e. able to grow along the root system of the crop plants) has been hypothesized. They could be applied to the soil as biofertilizers together with the crop, persist seasonally and avoid signal recognition by the seeds of the parasitic plants, thus preventing parasite seed germination and successive attachments to the host root (Boari et al. 2016). So far, these control methods have only been investigated in lab experiments and are thus, still far from practical field application.

3.5.4 Host Tolerance Through Low-SL Exudation

As discussed above SLs are the main germination stimulants for root parasitic plants. In studies that evaluated the induction of parasitic plant seed germination by exudates from different genotypes and cultivars of several crop species, a positive correlation was demonstrated between the SL concentration in the root exudate and the germination rate (Jamil et al. 2011a; Fernández-Aparicio et al. 2014; Yoneyama et al. 2015; Mohemed et al. 2018).

An approach for the management of parasitic plants in agricultural crops could thus be to reduce germination of the parasitic plant seeds by reducing the exudation of the germination stimulants. Several studies explored natural variation in germination stimulant production, for example, in the New Rice for Africa (NERICA) rice cultivars. This work showed that several NERICA cultivars (1, 2, 5, 10 and 17) displayed post-germination resistance to *S. hermonthica* and *S. asiatica* unlike

NERICA 7, 8 and 11, which were susceptible (Rodenburg et al. 2015). In parallel, variation in SL production in the NERICA genotypes was demonstrated, resulting in differences in *Striga* germination induction (Jamil et al. 2011b). The combination, by breeding, of germination-related resistance with post-germination resistance could result in better durable *Striga* resistance (Cissoko et al. 2011; Jamil et al. 2011b).

Also in pea there is evidence for a relationship between the total amount of SLs exuded and sensitivity to *O. crenata* infection (Pavan et al. 2016). A genotype with reduced SL exudation displayed partial field resistance. Evidence about this positive correlation does exist in sorghum as well. For example, sorghum-resistant genotype SRN39 produced much less 5-deoxystrigol than the susceptible Tabat (Yoneyama et al. 2010). Later, SRN39 proved to produce more orobanchol, instead of 5-deoxystrigol—due to a tentative modification in the SL biosynthetic pathway (Gobena et al. 2017). This mechanism occurs broader than just in SRN39 and was also observed in a number of other *Striga*-resistant sorghum genotypes (Mohemed et al. 2018). The same can be observed in maize where a modification in the SL composition seems to cause resistance. The *Striga*-susceptible cultivar (Pioneer 3253) produced mostly 5-deoxystrigol, whereas the *Striga*-resistant (KST 94) produced mostly sorgomol. Interestingly, the differences in SL composition in maize and sorghum did not affect the level of AM colonization (Yoneyama et al. 2015; Gobena et al. 2017).

Aside of exploiting natural variation, biotechnological approaches aiming to generate low SL exuding plants could be a strategy to reduce infestation by parasitic plants (López-Ráez et al. 2008). Indeed, it was demonstrated that tomato in which SL biosynthesis was knocked down through genetic modification was more resistant to *O. ramosa* infection (Kohlen et al. 2012). A reduction in SL production to obtain parasitic weed resistance was also achieved unintentionally. Dor and co-workers by using fast-neutron mutagenesis developed a tomato mutant (Sl-ORT1) resistant to various broomrape species (Dor et al. 2010). The Sl-ORT1 tomato was then discovered to be a SL-deficient mutant, and the resistance was thus associated to the low amount of strigolactones exuded (Dor et al. 2010). Breeding—through conventional or biotechnological approaches—for a reduction in the SL amount in exudates potentially also has negative consequences given their importance for the control of shoot and root architecture and the acquisition of nutrients through AM fungi (López-Ráez et al. 2008). This could possibly be prevented by approaches that reduce transport of SLs into the rhizosphere, which is facilitated by an ABC transporter, PDR1 (Borghi et al. 2015). However, under certain abiotic stress conditions, this could still negatively affect the adaptive capacity of plants by hampering AM fungi colonization. Particularly the example of sorghum and of maize shows that solutions in which the composition rather than the level of the SLs is changed may be the best solution (Yoneyama et al. 2015; Gobena et al. 2017). Nevertheless, several examples show that a reduction in SL production results in an acceptable level of resistance without large consequences for the plant phenotype (Jamil et al. 2011a; Pavan et al. 2016). In order to prevent that this partial germination-based resistance is overcome, a combination of pre- and post-attachment resistance mechanisms is necessary.

3.5.5 Parasitic Plant Seed Germination Bioassay

Plant seeds germinate when they are exposed to appropriate temperature, humidity, oxygen and, often, light. In case of seeds of parasitic plants, these conditions except light are also required, but SLs have a pivotal role in the regulation of germination. As said above, seeds of these root parasites will only germinate if they perceive the presence of the stimulants, which in the field means they are within the host rhizosphere and thus after germination they have a better chance to rapidly attach to the host root.

Considering the extreme biotic and abiotic complexity of the rhizosphere, a simple bioassay for studying SLs has been used extensively since the discovery of the stimulating compounds. This assay (Mangnus et al. 1992), with a number of adaptations and variants, is based on the reproduction, in vitro, of the steps necessary for parasitic seeds to germinate. Thus, seeds are first kept in a moist environment (i.e. on wet paper discs in Petri dishes), at a constant temperature (around 22–25 °C), for some days depending on the species. This mimics the so-called conditioning phase (see above). After that, seeds are placed in contacts with the stimulant in a proper concentration (usually at ppm or ppb levels) in order to induce germination. This happens a few days after stimulant application. Several observations can be then performed, e.g. percentage of germination, shape and length of the germination tubes, seed viability. More recently, high-throughput germination bioassays have been developed based on a standardized 96-well plate test coupled with spectrophotometric reading of tetrazolium salt (MTT) reduction (Pouvreau et al. 2013). These bioassays can be useful for different purposes, e.g. to guide the purification steps for the identification of novel stimulants; to test dose-response effectiveness of SLs, derivatives and analogues; to evaluate SL selectivity/specificity in parasitic species/strains; to bioassay germination inhibitors; and to study the physiology of the first stages of the parasitism.

3.6 Prospects

The research on SLs and parasitic plants has received an enormous attention in the last one to two decades, both because of the discovery of other important roles of the SLs (see above), and because of the extraordinary technological progresses, which made available equipment and tools unimaginable just a few years ago. High-throughput bioassays allow a faster and more accurate evaluation of the compound bioactivity, purification and analytical procedures, and structure determination has been simplified by more sophisticated, sensitive and automated equipment; "omics" approaches allow an easier understanding of the mechanisms of action of stimulants and inhibitors. Considering the key role of SLs in plant parasitism, parasitic weed management strategies should be developed in this perspective. Indeed, the level of success in controlling these parasites is very often still inadequate. The factors influencing the parasitic weed cycle have not yet been completely deciphered and thus the capability of predicting their infectiveness and infestation is still limited. The only option for success in such a difficult field of research is to bring together scientists representing a wide spectrum of disciplines, advanced research approaches and geographical representation of parasitic plant research. Assembling specialists with different perspectives, all focused around the common theme of plant parasitism, could provide a stimulating opportunity for finding widely usable, novel strategies for parasitic weed management.

Glossary

ABC transporter (ATP-binding cassette transporter) Transport protein, consisting of a transmembrane domain and membrane-associated ATPase, that utilizes the energy of ATP to transport substrates across cellular membranes.

Allelopathy The phenomenon that plants release molecules (called allelochemicals) that affect seed germination, plant physiology, growth and survival of other plants.

Arbuscular mycorrhizal (AM) fungi A group of obligate fungal root biotrophs that engage in symbiosis with 80% of all land plants. They penetrate the cortical cells of the roots of a vascular plant, forming unique structures, arbuscules, that help plants to capture nutrients such as phosphorus, sulphur, nitrogen and micronutrients from the soil and get photoassimilates of the plant in return.

Aromatic ring A cyclic (ring-shaped), planar (flat) molecule with a ring of resonance bonds that confers high stability to the molecule. The simplest aromatic compound is benzene, and the most common aromatic compounds are derived from it.

Bioactiphore The active part of a molecule responsible for the biological activity of the compound.

Biofuel Fuel derived directly from plants or indirectly from agricultural, commercial, domestic and/or industrial waste.

Carotenoids Organic pigments produced by plants and algae, in which they play an important role as accessory pigments in photosynthesis, as well as by several bacteria and fungi. Carotenoids are also precursors for cell signalling molecules, e.g. abscisic acid, which regulates plant growth, seed dormancy, embryo maturation and germination, cell division and elongation, floral growth and stress responses.

Seed dormancy A process that prevents germination of an intact viable seed in a specified period of time under any combination of normal physical environmental factors that are otherwise favourable for its germination.

Gene silencing Interruption or suppression of the expression of a gene at the transcriptional or translational level.

Intercrop A crop grown between the rows of another crop.

Isomer A molecule with the same molecular formula as another molecule but with a different chemical structure.

Nodulation The process of forming root nodules containing symbiotic, nitrogen fixing and bacteria.

Noncanonical SLs SLs lacking the A, B or C ring but still retaining the enol ether-D ring moiety, which is essential for biological activity.

Nucleophilic agent A reagent that forms a bond to its reaction partner (the electrophile) by donating both bonding electrons.

Phloem The living tissue that transports the soluble organic compounds made in the leaves during photosynthesis to all other parts of the plant.

Rhizosphere The zone of soil surrounding a plant root where the biology and chemistry of the soil are directly affected by a plant's root system, associated root secretions and microorganisms.

Xylem *Plant* vascular tissue that conveys water and dissolved minerals from the roots to the rest of the plant and also provides physical support.

References

Abbes Z, Kharrat M, Delavault P et al (2007) Field evaluation of the resistance of some faba bean (*Vicia faba* L.) genotypes to the parasitic weed *Orobanche foetida* Poiret. Crop Prot 26:1777–1784. https://doi.org/10.1016/j.cropro.2007.03.012

Abdel-Lateif K, Bogusz D, Hocher V (2012) The role of flavonoids in the establishment of plant roots endosymbioses with arbuscular mycorrhiza fungi, rhizobia and Frankia bacteria. Plant Signal Behav 7:636–641. https://doi.org/10.4161/psb.20039

Abebe G, Sahile G, Abdel-Rahman MA-T (2005) Effect of soil solarization on *Orobanche* soil seed bank and tomato yield in Central Rift Valley of Ethiopia. World J Agric Sci 1:143–147

Acharya BD (2014) Assessment of different non-host crops as trap crop for reducing *Orobanche aegyptiaca* Pers. seed bank. Ecoprint An Int J Ecol 19:31–38. https://doi.org/10.3126/eco.v19i0.9851

Acharya BD, Khattri GB, Chettri MK, Srivastava SC (2002) Effect of *Brassica campestris* var. *toria* as a catch crop on *Orobanche aegyptiaca* seed bank. Crop Prot 21:533–537. https://doi.org/10.1016/S0261-2194(01)00137-5

Akiyama K, Matsuzaki KI, Hayashi H (2005) Plant sesquiterpenes induce hyphal branching in arbuscular mycorrhizal fungi. Nature 435:824–827. https://doi.org/10.1038/nature03608

Aksoy E, Arslan ZF, Tetik ES (2015) Using the possibilities of some trap, catch and Brassicaceaen crops for controlling crenate broomrape a problem in lentil fields. Int J Plant Prod 10:53–62

Al-Menoufi OA (1989) Crop rotation as a control measure of *Orobanche crenata* in Vicia faba fields. In: Wegmann K, Musselman L (eds) Progress in *Orobanche* research. Eberhard-Karl-Universitat, Tubingen, pp 241–247

Aly R, Dubey NK, Yahyaa M et al (2014) Gene silencing of CCD7 and CCD8 in *Phelipanche aegyptiaca* by tobacco rattle virus system retarded the parasite development on the host. Plant Signal Behav 9:e29376. https://doi.org/10.4161/psb.29376

Auger B, Pouvreau JB, Pouponneau K et al (2012) Germination stimulants of *Phelipanche ramosa* in the rhizosphere of *Brassica napus* are derived from the glucosinolate pathway. Mol Plant Microbe Interact 25:993–1004. https://doi.org/10.1094/MPMI-01-12-0006-R

Babiker AG, Hamdoun AM (1982) Factors affecting the activity of GR7 in stimulating germination of *Striga hermonthica* (Del.) Benth. Weed Res 22:111–115. https://doi.org/10.1111/j.1365-3180.1982.tb00152.x

Babiker AG, Ahmed E, Dawoud D, Abdella N (2007) *Orobanche* species in Sudan: history, distribution and management. Sudan J Agric Res 10:107–114

Bais HP, Weir TL, Perry LG et al (2006) The role of root exudates in rhizosphere interactions with plants and other organisms. Annu Rev Plant Biol 57:233–266. https://doi.org/10.1146/annurev.arplant.57.032905.105159

Bar-Nun N, Sachs T, Mayer AM (2008) A role for IAA in the infection of *Arabidopsis thaliana* by *Orobanche aegyptiaca*. Ann Bot 101:261–265. https://doi.org/10.1093/aob/mcm032

Bedi JS, Kapur SP, Mohan C (1997) *Orobanche* – a threat to raya and taramira in Punjab. J Res (Punjab Agric Univ) 34:149–152

Besserer A, Puech-Pagès V, Kiefer P et al (2006) Strigolactones stimulate arbuscular mycorrhizal fungi by activating mitochondria. PLoS Biol 4:e226. https://doi.org/10.1371/journal.pbio.0040226

Boari A, Ciasca B, Pineda-Martos R et al (2016) Parasitic weed management by using strigolactones-degrading fungi. Pest Manag Sci 72:2043–2047. https://doi.org/10.1002/ps.4226

Borghi L, Kang J, Ko D et al (2015) The role of ABCG-type ABC transporters in phytohormone transport. Biochem Soc Trans 43:924–930. https://doi.org/10.1042/BST20150106

Bouwmeester HJ, Matusova R, Zhongkui S, Beale MH (2003) Secondary metabolite signalling in host–parasitic plant interactions. Curr Opin Plant Biol 6:358–364. https://doi.org/10.1016/S1369-5266(03)00065-7

Bouwmeester HJ, Roux C, López-Ráez JA, Bécard G (2007) Rhizosphere communication of plants, parasitic plants and AM fungi. Trends Plant Sci 12:224–230. https://doi.org/10.1016/j.tplants.2007.03.009

Bülbül F, Aksoy E, Uygur S, Uygur N (2009) Broomrape (*Orobanche* spp.) problem in the eastern mediterranean region of Turkey. Helia 32:141–152. https://doi.org/10.2298/HEL0951141B

Cagáň L, Tóth P (2003) A decrease in tomato yield caused by branched broomrape (*Orobanche ramosa*) parasitization. Acta Fytotech Zootech 6:65–68

Cardoso C, Zhang Y, Jamil M et al (2014) Natural variation of rice strigolactone biosynthesis is associated with the deletion of two MAX1 orthologs. Proc Natl Acad Sci U S A 111:2379–2384. https://doi.org/10.1073/pnas.1317360111

Carson AG (1989) Effect of intercropping sorghum and groundnuts on density of *Striga hermonthica* in The Gambia. Trop Pest Manag 35:130–132. https://doi.org/10.1080/09670878909371340

Chang M, Lynr DG, Netzly DH, Butler LG (1986) Chemical regulation of distance: characterization of the first natural host germination stimulant for *Striga asiatica*. J Am Chem Soc 108:7858–7860. https://doi.org/10.1021/ja00284a074

Charnikhova TV, Gaus K, Lumbroso A et al (2017) Zealactones. Novel natural strigolactones from maize. Phytochemistry 137:123–131. https://doi.org/10.1016/j.phytochem.2017.02.010

Charnikhova TV, Gaus K, Lumbroso A et al (2018) Zeapyranolactone – a novel strigolactone from maize. Phytochem Lett 24:172–178. https://doi.org/10.1016/j.phytol.2018.01.003

Cheng X, Flokova K, Bouwmeester H, Ruyter-Spira C (2017) The role of endogenous strigolactones and their interaction with ABA during the infection process of the parasitic weed *Phelipanche ramosa* in tomato plants. Front Plant Sci 8:392. https://doi.org/10.3389/fpls.2017.00392

Chiou TJ, Lin SI (2011) Signaling network in sensing phosphate availability in plants. Annu Rev Plant Biol 62:185–206. https://doi.org/10.1146/annurev-arplant-042110-103849

Cissoko M, Boisnard A, Rodenburg J et al (2011) New Rice for Africa (NERICA) cultivars exhibit different levels of post-attachment resistance against the parasitic weeds *Striga hermonthica* and *Striga asiatica*. New Phytol 192:952–963. https://doi.org/10.1111/j.1469-8137.2011.03846.x

Das M, Fernández-Aparicio M, Yang Z et al (2015) Parasitic plants *Striga* and *Phelipanche* dependent upon exogenous strigolactones for germination have retained genes for strigolactone biosynthesis. Am J Plant Sci 6:1151–1166. https://doi.org/10.4236/ajps.2015.68120

De Groote H, Wangare L, Kanampiu F et al (2008) The potential of a herbicide resistant maize technology for *Striga* control in Africa. Agric Syst 97:83–94. https://doi.org/10.1016/j.agsy.2007.12.003

Dhanapal GN, Struik PC, Udayakumar M, Timmermans PCJM (1996) Management of broomrape (*Orobanche* spp.) – a review. J Agron Crop Sci 176:335–359. https://doi.org/10.1111/j.1439-037X.1996.tb00479.x

Díaz JS, Norambuena H, López-Granados FM (2006) Characterization of the holoparasitism of *Orobanche ramosa* on tomatoes under field conditions. Agric Téc 66:223–234

Dita MA, Die JV, Román B et al (2009) Gene expression profiling of *Medicago truncatula* roots in response to the parasitic plant *Orobanche crenata*. Weed Res 49:66–80. https://doi.org/10.1111/j.1365-3180.2009.00746.x

Domagalska MA, Leyser O (2011) Signal integration in the control of shoot branching. Nat Rev Mol Cell Biol 12:211–221. https://doi.org/10.1038/nrm3088

Dor E, Alperin B, Wininger S et al (2010) Characterization of a novel tomato mutant resistant to the weedy parasites *Orobanche* and *Phelipanche* spp. Euphytica 171:371–380. https://doi.org/10.1007/s10681-009-0041-2

Dugje IY, Kamara AY, Omoigui LO (2006) Infestation of crop fields by *Striga* species in the savanna zones of Northeast Nigeria. Agric Ecosyst Environ 116:251–254. https://doi.org/10.1016/j.agee.2006.02.013

Eizenberg H, Golan S, Joel DM (2002) First report of the parasitic plant *Orobanche aegyptiaca* infecting olive. Plant Dis 86:814. https://doi.org/10.1094/PDIS.2002.86.7.814A

Ejeta G (2007) The Striga scourge in Africa: a growing pandemic. In: Integrating new technologies for *Striga* control. World Scientific, Singapore, pp 3–16

Emechebe AM, Singh BB, Leleji OI, et al (1991) Cowpea-striga problems and research in Nigeria. In: Combating striga in Africa: proceedings of the international workshop held in Ibadan, Nigeria, 22–24 Aug 1988. International Institute of Tropical Agriculture, pp 18–28

Evidente A, Fernández-Aparicio M, Cimmino A et al (2009) Peagol and peagoldione, two new strigolactone-like metabolites isolated from pea root exudates. Tetrahedron Lett 50:6955–6958. https://doi.org/10.1016/j.tetlet.2009.09.142

Evidente A, Cimmino A, Fernández-Aparicio M et al (2011) Soyasapogenol B and trans-22-dehydrocampesterol from common vetch (*Vicia sativa* L.) root exudates stimulate broomrape seed germination. Pest Manag Sci 67:1015–1022. https://doi.org/10.1002/ps.2153

Fernández-Aparicio M, Emeran AA, Rubiales D (2008) Control of *Orobanche crenata* in legumes intercropped with fenugreek (*Trigonella foenum-graecum*). Crop Prot 27:653–659. https://doi.org/10.1016/j.cropro.2007.09.009

Fernández-Aparicio M, Emeran AA, Rubiales D (2010) Inter-cropping with berseem clover (*Trifolium alexandrinum*) reduces infection by *Orobanche crenata* in legumes. Crop Prot 29:867–871. https://doi.org/10.1016/j.cropro.2010.03.004

Fernández-Aparicio M, Kisugi T, Xie X et al (2014) Low strigolactone root exudation: a novel mechanism of broomrape (*Orobanche* and *Phelipanche* spp.) resistance available for faba bean breeding. J Agric Food Chem 62:7063–7071. https://doi.org/10.1021/jf5027235

Foo E, Yoneyama K, Hugill CJ et al (2013) Strigolactones and the regulation of pea symbioses in response to nitrate and phosphate deficiency. Mol Plant 6:76–87. https://doi.org/10.1093/mp/sss115

Gbèhounou G, Adango E (2003) Trap crops of *Striga hermonthica*: *in vitro* identification and effectiveness in situ. Crop Prot 22:395–404. https://doi.org/10.1016/S0261-2194(02)00196-5

Gibot-Leclerc S (2003) Rôle potentiel des plantes adventices du colza d'hiver dans l'extension de l'orobanche rameuse en Poitou-Charentes (Potential role of winter rape weeds in the extension of broomrape in Poitou-Charentes). C R Biol 326:645–658. https://doi.org/10.1016/S1631-0691(03)00169-0

Gibot-Leclerc S, Perronne R, Dessaint F et al (2016) Assessment of phylogenetic signal in the germination ability of *Phelipanche ramosa* on Brassicaceae hosts. Weed Res 56:452–461. https://doi.org/10.1111/wre.12222

Gobena D, Shimels M, Rich PJ et al (2017) Mutation in sorghum LOW GERMINATION STIMULANT 1 alters strigolactones and causes *Striga* resistance. Proc Natl Acad Sci U S A 114:4471–4476. https://doi.org/10.1073/pnas.1618965114

Gressel J, Hanafi A, Head G et al (2004) Major heretofore intractable biotic constraints to African food security that may be amenable to novel biotechnological solutions. Crop Prot 23:661–689. https://doi.org/10.1016/j.cropro.2003.11.014

Haq BU, Ahmad MZ, Ur Rehman N et al (2017) Functional characterization of soybean strigolactone biosynthesis and signaling genes in Arabidopsis MAX mutants and GmMAX3 in soybean nodulation. BMC Plant Biol 17:259. https://doi.org/10.1186/s12870-017-1182-4

Hooper AM, Tsanuo MK, Chamberlain K et al (2010) Isoschaftoside, a C-glycosylflavonoid from *Desmodium uncinatum* root exudate, is an allelochemical against the development of *Striga*. Phytochemistry 71:904–908. https://doi.org/10.1016/j.phytochem.2010.02.015

Ito S, Ito K, Abeta N et al (2016) Effects of strigolactone signaling on *Arabidopsis* growth under nitrogen deficient stress condition. Plant Signal Behav 11:e1126031. https://doi.org/10.1080/15592324.2015.1126031

Jamil M, Charnikhova T, Cardoso C et al (2011a) Quantification of the relationship between strigolactones and *Striga hermonthica* infection in rice under varying levels of nitrogen and phosphorus. Weed Res 51:373–385. https://doi.org/10.1111/j.1365-3180.2011.00847.x

Jamil M, Rodenburg J, Charnikhova T, Bouwmeester HJ (2011b) Pre-attachment *Striga hermonthica* resistance of New Rice for Africa (NERICA) cultivars based on low strigolactone production. New Phytol 192:964–975. https://doi.org/10.1111/j.1469-8137.2011.03850.x

Jamil M, Charnikhova T, Houshyani B et al (2012) Genetic variation in strigolactone production and tillering in rice and its effect on *Striga hermonthica* infection. Planta 235:473–484. https://doi.org/10.1007/s00425-011-1520-y

Jamil M, Van Mourik TA, Charnikhova T, Bouwmeester HJ (2013) Effect of diammonium phosphate application on strigolactone production and *Striga hermonthica* infection in three sorghum cultivars. Weed Res 53:121–130. https://doi.org/10.1111/wre.12003

Jamil M, Charnikhova T, Jamil T et al (2014a) Influence of fertilizer microdosing on strigolactone production and *Striga hermonthica* parasitism in pearl millet. Int J Agric Biol 16:935–940

Jamil M, Charnikhova T, Verstappen F et al (2014b) Effect of phosphate-based seed priming on strigolactone production and *Striga hermonthica* infection in cereals. Weed Res 54:307–313. https://doi.org/10.1111/wre.12067

Joel DM, Hershenhorn J, Eizenberg H et al (2007) Biology and management of weedy root parasites. In: Janick J (ed) Horticultural reviews. Wiley, London, pp 267–350

Joel DM, Chaudhuri SK, Plakhine D et al (2011) Dehydrocostus lactone is exuded from sunflower roots and stimulates germination of the root parasite *Orobanche cumana*. Phytochemistry 72:624–634. https://doi.org/10.1016/j.phytochem.2011.01.037

Johnson DE, Riches CR, Diallo R, Jones MJ (1997) *Striga* on rice in West Africa; crop host range and the potential of host resistance. Crop Prot 16:153–157. https://doi.org/10.1016/S0261-2194(96)00079-8

Kannan C, Zwanenburg B (2014) A novel concept for the control of parasitic weeds by decomposing germination stimulants prior to action. Crop Prot 61:11–15. https://doi.org/10.1016/j.cropro.2014.03.008

Kgosi RL, Zwanenburg B, Mwakaboko AS, Murdoch AJ (2012) Strigolactone analogues induce suicidal seed germination of *Striga* spp. in soil. Weed Res 52:197–203. https://doi.org/10.1111/j.1365-3180.2012.00912.x

Khan ZR, Midega CAO, Bruce TJA et al (2010) Exploiting phytochemicals for developing a 'push-pull' crop protection strategy for cereal farmers in Africa. J Exp Bot 61:4185–4196. https://doi.org/10.1093/jxb/erq229

Kim SK, Akintunde AY, Walker P (1994) Responses of maize, sorghum and millet host plants to infestation by *Striga hermonthica*. Crop Prot 13:582–590. https://doi.org/10.1016/0261-2194(94)90003-5

Kleifeld Y, Goldwasser Y, Herlzlinger G et al (1994) The effects of flax (*Linum usitatissimum* L.) and other crops as trap and catch crops for control of Egyptian broomrape (*Orobanche aegyptiaca* Pers.). Weed Res 34:37–44. https://doi.org/10.1111/j.1365-3180.1994.tb01971.x

Kohlen W, Charnikhova T, Lammers M et al (2012) The tomato *CAROTENOID CLEAVAGE DIOXYGENASE8* (*SlCCD8*) regulates rhizosphere signaling, plant architecture and affects reproductive development through strigolactone biosynthesis. New Phytol 196:535–547. https://doi.org/10.1111/j.1469-8137.2012.04265.x

Koltai H (2011) Strigolactones are regulators of root development. New Phytol 190:545–549. https://doi.org/10.1111/j.1469-8137.2011.03678.x

Kumar M, Pandya-Kumar N, Kapulnik Y, Koltai H (2015) Strigolactone signaling in root development and phosphate starvation. Plant Signal Behav 10:e1045174. https://doi.org/10.1080/15592324.2015.1045174

Labrada R (2007) Progress on farmers training on parasitic weed management. Food Agriculture Organisation United Nations, p 156

Lei L (2017) Parasitic plants: injecting hormone into host. Nat Plants 3:17084. https://doi.org/10.1038/nplants.2017.84

Letousey P, De Zélicourt A, Vieira Dos Santos C et al (2007) Molecular analysis of resistance mechanisms to *Orobanche cumana* in sunflower. Plant Pathol 56:536–546. https://doi.org/10.1111/j.1365-3059.2007.01575.x

Lins RD, Colquhoun JB, Mallory-Smith CA (2006) Investigation of wheat as a trap crop for control of *Orobanche minor*. Weed Res 46:313–318. https://doi.org/10.1111/j.1365-3180.2006.00515.x

Liu CW, Murray JD (2016) The role of flavonoids in nodulation host-range specificity: an update. Plants (Basel) 5:(3)33. https://doi.org/10.3390/plants5030033

Liu Q, Zhang Y, Matusova R et al (2014) *Striga hermonthica* MAX2 restores branching but not the very low fluence response in the *Arabidopsis thaliana* max2 mutant. New Phytol 202:531–541. https://doi.org/10.1111/nph.12692

López-Ráez JA, Matusova R, Cardoso C et al (2008) Strigolactones: ecological significance and use as a target for parasitic plant control. Pest Manag Sci 64:471–477. https://doi.org/10.1002/ps.1692

López-Ráez JA, Charnikhova T, Gomez-Roldan V et al (2008) Tomato strigolactones are derived from carotenoids and their biosynthesis is promoted by phosphate starvation. New Phytol 178:863–874. https://doi.org/10.1111/j.1469-8137.2008.02406.x

Losner-Goshen D, Portnoy VH, Mayer AM, Joel DM (1998) Pectolytic activity by the haustorium of the parasitic plant *Orobanche* L. (Orobanchaceae) in host roots. Ann Bot 81:319–326. https://doi.org/10.1006/anbo.1997.0563

Mallory-Smith C, Colquhoun J (2012) Small broomrape (*Orobanche minor*) in Oregon and the 3 Rs: regulation, research, and reality. Weed Sci 60:277–282. https://doi.org/10.1614/WS-D-11-00078.1

Mangnus EM, Stommen PLA, Zwanenburg B (1992) A standardized bioassay for evaluation of potential germination stimulants for seeds of parasitic weeds. J Plant Growth Regul 11:91–98. https://doi.org/10.1007/BF00198020

Manyong VM, Alene AD, Olanrewaju A et al (2007) Baseline study of *Striga* control using IR maize in Western Kenya. AATF/IITA Striga Control Project, pp 27–31

Marzec M, Muszynska A, Gruszka D (2013) The role of strigolactones in nutrient-stress responses in plants. Int J Mol Sci 14:9286–9304. https://doi.org/10.3390/ijms14059286

Matusova R, van Mourik T, Bouwmeester HJ (2004) Changes in the sensitivity of parasitic weed seeds to germination stimulants. Seed Sci Res 14:335–344. https://doi.org/10.1079/SSR2004187

Matusova R, Rani K, Verstappen FW et al (2005) The strigolactone germination stimulants of the plant-parasitic *Striga* and *Orobanche* spp. are derived from the carotenoid pathway. Plant Physiol 139:920–934. https://doi.org/10.1104/pp.105.061382

Mohamed KI, Musselman LJ, Riches CR (2001) The genus *Striga* (Scrophulariaceae) in Africa. Ann Mo Bot Gard 88:60–103. https://doi.org/10.2307/2666132

Mohemed N, Charnikhova T, Fradin EF et al (2018) Genetic variation in Sorghum bicolor strigolactones and their role in resistance against *Striga hermonthica*. J Exp Bot 69:2415–2430. https://doi.org/10.1093/jxb/ery041

Motazedi S, Jahedi A, Farnia A (2010) Integrated broomrape (*Orobanche aegyptiaca*) control by sulfosulfuron (WG 75%) herbicide with wheat mulch applied in field potato. In: Proceedings of 3rd Iranian weed science congress, volume 2: key papers, weed management and herbicides, Babolsar, Iran, 17–18 Feb 2010. Iranian Society of Weed Science, Tehran, pp 227–229

Musselman LJ, Parker C (1982) Preliminary host ranges of some strains of economically important broomrapes (*Orobanche*). Econ Bot 36:270–273. https://doi.org/10.1007/BF02858547

Neondo JO, Alakonya AE, Kasili RW (2017) Screening for potential *Striga hermonthica* fungal and bacterial biocontrol agents from suppressive soils in Western Kenya. BioControl 62:705–717. https://doi.org/10.1007/s10526-017-9833-9

Oswald A, Ransom JK (2001) *Striga* control and improved farm productivity using crop rotation. Crop Prot 20:113–120. https://doi.org/10.1016/S0261-2194(00)00063-6

Parker C (1994) The present state of the *Orobanche* problem. In: Pieterse AH, Verkleij JAC, ter Borg SJ (eds) Biology and management of *Orobanche*. Proceedings of the third international workshop on *Orobanche* and related *Striga* research, Amsterdam, Netherlands, 8–12 Nov 1993. Royal Tropical Institute, Amsterdam, pp 17–26

Parker C (2009) Observations on the current status of *Orobanche* and *Striga* problems worldwide. Pest Manag Sci 65:453–459. https://doi.org/10.1002/ps.1713

Parker C (2013) The parasitic weeds of the Orobanchaceae. In: Joel DM, Gressel J, Musselman LJ (eds) Parasitic Orobanchaceae: parasitic mechanisms and control strategies. Springer, Berlin, pp 313–344

Parker C, Riches CR (1993) Parasitic weeds of the world: biology and control. CAB International, Wallingford

Pavan S, Schiavulli A, Marcotrigiano AR et al (2016) Characterization of low-strigolactone germplasm in pea (*Pisum sativum* L.) resistant to crenate broomrape (*Orobanche crenata* Forsk.). Mol Plant Microbe Interact 29:743–749. https://doi.org/10.1094/MPMI-07-16-0134-R

Péret B, Clément M, Nussaume L, Desnos T (2011) Root developmental adaptation to phosphate starvation: better safe than sorry. Trends Plant Sci 16:442–450. https://doi.org/10.1016/j.tplants.2011.05.006

Pérez-de-Luque A, Galindo JC, Macías FA, Jorrin J (2000) Sunflower sesquiterpene lactone models induce *Orobanche cumana* seed germination. Phytochemistry 53:45–50. https://doi.org/10.1016/S0031-9422(99)00485-9

Perronne R, Gibot-Leclerc S, Dessaint F et al (2017) Is induction ability of seed germination of *Phelipanche ramosa* phylogenetically structured among hosts? A case study on Fabaceae species. Genetica 145:481–489. https://doi.org/10.1007/s10709-017-9990-x

Pickett JA, Hamilton ML, Hooper AM et al (2010) Companion cropping to manage parasitic plants. Annu Rev Phytopathol 48:161–177. https://doi.org/10.1146/annurev-phyto-073009

Pouvreau JB, Gaudin Z, Auger B et al (2013) A high-throughput seed germination assay for root parasitic plants. Plant Methods 9:32. https://doi.org/10.1186/1746-4811-9-32

Press MC, Shah N, Tuohy JM, Stewart GR (1987) Carbon isotope ratios demonstrate carbon flux from C4 host to C3 parasite. Plant Physiol 85:1143–1145. https://doi.org/10.1104/pp.85.4.1143

Riopel JL, Timko MP (1995) Haustorial initiation and differentiation. In: Press MC, Graves JD (eds) Parasitic plants. Chapman & Hall, London, pp 39–79

Rodenburg J, Bastiaans L, Weltzien E, Hess DE (2005) How can field selection for *Striga* resistance and tolerance in sorghum be improved? Field Crops Res 93:34–50. https://doi.org/10.1016/j.fcr. 2004.09.004

Rodenburg J, Cissoko M, Kayeke J et al (2015) Do NERICA rice cultivars express resistance to *Striga hermonthica* (Del.) Benth. and *Striga asiatica* (L.) Kuntze under field conditions? Field Crops Res 170:83–94. https://doi.org/10.1016/j.fcr.2014.10.010

Román B, Satovic Z, Alfaro C et al (2007) Host differentiation in *Orobanche foetida* Poir. Flora Morphol Distrib Funct Ecol Plants 202:201–208. https://doi.org/10.1016/j.flora.2006.07.003

Rubiales D, Fernández-Aparicio M, Wegmann K, Joel DM (2009) Revisiting strategies for reducing the seedbank of *Orobanche* and *Phelipanche* spp. Weed Res 49:23–33. https://doi.org/10. 1111/j.1365-3180.2009.00742.x

Ruyter-Spira C, Kohlen W, Charnikhova T et al (2011) Physiological effects of the synthetic strigolactone analog GR24 on root system architecture in *Arabidopsis*: another belowground role for strigolactones? Plant Physiol 155:721–734. https://doi.org/10.1104/pp.110.166645

Ruyter-Spira C, Al-Babili S, van der Krol S, Bouwmeester H (2013) The biology of strigolactones. Trends Plant Sci 18:72–83. https://doi.org/10.1016/j.tplants.2012.10.003

Samejima H, Babiker AG, Takikawa H et al (2016) Practicality of the suicidal germination approach for controlling *Striga hermonthica*. Pest Manag Sci 72:2035–2042. https://doi.org/ 10.1002/ps.4215

Sauerborn J (1991) The economic importance of the phytoparasites *Orobanche* and *Striga*. In: Ransom JK, Musselman LJ, Worsham AD, Parker C (eds) Proceedings of the 5th international symposium of parasitic weeds, Nairobi, Kenya, 24–30 June 1991. CIMMYT (International Maize and Wheat Improvement Center), Nairobi, pp 137–143

Sauerborn J, Saxena MC (1986) A review on agronomy in relation to *Orobanche* problems in faba bean (*Vicia faba* L.). In: ter Borg S (ed) Proceedings of a workshop on biology and control of *Orobanche*, Wageningen, Netherlands, 13–17 Jan 1986. Landbouwuniversiteit, pp 160–165

Saxena MC, Linke KH, Sauerborn J (1994) Integrated control of *Orobanche* in cool-season food legumes. In: Pieterse A, Verkleij J, Borg S (eds) Biology and management of *Orobanche*. Royal Tropical Institute, Amsterdam, pp 419–431

Schlemper TR, Leite MFA, Lucheta AR et al (2017) Rhizobacterial community structure differences among sorghum cultivars in different growth stages and soils. FEMS Microbiol Ecol 93. https://doi.org/10.1093/femsec/fix096

Scholes JD, Press MC (2008) *Striga* infestation of cereal crops – an unsolved problem in resource limited agriculture. Curr Opin Plant Biol 11:180–186. https://doi.org/10.1016/j.pbi.2008.02.004

Spallek T, Melnyk CW, Wakatake T et al (2017) Interspecies hormonal control of host root morphology by parasitic plants. Proc Natl Acad Sci U S A 114:5283–5288. https://doi.org/10. 1073/pnas.1619078114

Sun H, Tao J, Liu S et al (2014) Strigolactones are involved in phosphate- and nitrate-deficiency-induced root development and auxin transport in rice. J Exp Bot 65:6735–6746. https://doi.org/ 10.1093/jxb/eru029

Swanton CJ, Booth BD (2004) Management of weed seedbanks in the context of populations and communities. Weed Technol 18:1496–1502. https://doi.org/10.1614/0890-037X(2004)018[1496:MOWSIT]2.0.CO;2

Tasker AV, Westwood JH (2012) The U.S. witchweed eradication effort turns 50: a retrospective and look-ahead on parasitic weed management. Weed Sci 60:267–268. https://doi.org/10.1614/ WS-D-12-00003.1

Těšitel J, Plavcová L, Cameron DD (2010) Interactions between hemiparasitic plants and their hosts: the importance of organic carbon transfer. Plant Signal Behav 5:1072–1076. https://doi. org/10.4161/psb.5.9.12563

Timus A, Croitoru N (2007) The state of tobacco culture in Republic Moldova and phytosanitary problems of tobacco production. Rasteniev'dni Nauk 44:209–212

Torres-Vera R, Garcia JM, Pozo MJ, López-Ráez JA (2014) Do strigolactones contribute to plant defence? Mol Plant Pathol 15:211–216. https://doi.org/10.1111/mpp.12074

Ueno K, Fujiwara M, Nomura S et al (2011a) Structural requirements of strigolactones for germination induction of *Striga gesnerioides* seeds. J Agric Food Chem 59:9226–9231. https://doi.org/10.1021/jf202418a

Ueno K, Nomura S, Muranaka S et al (2011b) Ent-2′-epi-orobanchol and its acetate, as germination stimulants for *Striga gesnerioides* seeds isolated from cowpea and red clover. J Agric Food Chem 59:10485–10490. https://doi.org/10.1021/jf2024193

Ueno K, Furumoto T, Umeda S et al (2014) Heliolactone, a non-sesquiterpene lactone germination stimulant for root parasitic weeds from sunflower. Phytochemistry 108:122–128. https://doi.org/10.1016/j.phytochem.2014.09.018

Wang Y, Bouwmeester HJ (2018) Structural diversity in the strigolactones. J Exp Bot 69:2219–2230. https://doi.org/10.1093/jxb/ery091

Warren P (2006) The branched broomrape eradication program in Australia. In: 15th Australian weeds conference, South Australia, managing weeds in a changing climate. Weed Management Society of South Australia, Adelaide, pp 610–613

Westwood JH (2013) The physiology of the established parasite–host association. In: Joel DM, Gressel J, Musselman LJ (eds) Parasitic Orobanchaceae. Springer, Berlin, pp 87–114

Yoneyama K, Awad AA, Xie X et al (2010) Strigolactones as germination stimulants for root parasitic plants. Plant Cell Physiol 51:1095–1103. https://doi.org/10.1093/pcp/pcq055

Yoneyama K, Xie X, Kim HI et al (2012) How do nitrogen and phosphorus deficiencies affect strigolactone production and exudation? Planta 235:1197–1207. https://doi.org/10.1007/s00425-011-1568-8

Yoneyama K, Arakawa R, Ishimoto K et al (2015) Difference in *Striga*-susceptibility is reflected in strigolactone secretion profile, but not in compatibility and host preference in arbuscular mycorrhizal symbiosis in two maize cultivars. New Phytol 206:983–989. https://doi.org/10.1111/nph.13375

Zonno MC, Montemurro P, Vurro M (2000) *Orobanche ramosa*, un'infestante parassita in espansione nell'Italia meridionale. Inf Fitopatol 4:13–21

Zwanenburg B, Pospíšil T (2013) Structure and activity of strigolactones: new plant hormones with a rich future. Mol Plant 6:38–62. https://doi.org/10.1093/mp/sss141

Zwanenburg B, Mwakaboko AS, Reizelman A et al (2009) Structure and function of natural and synthetic signalling molecules in parasitic weed germination. Pest Manag Sci 65:478–491. https://doi.org/10.1002/ps.1706

Zwanenburg B, Ćavar Zeljković S, Pospíšil T (2016) Synthesis of strigolactones, a strategic account. Pest Manag Sci 72:15–29. https://doi.org/10.1002/ps.4105

Chapter 4
The Role of Strigolactones in Plant–Microbe Interactions

Soizic Rochange, Sofie Goormachtig, Juan Antonio Lopez-Raez, and Caroline Gutjahr

Abstract Plants associate with an infinite number of microorganisms that interact with their hosts in a mutualistic or parasitic manner. Evidence is accumulating that strigolactones (SLs) play a role in shaping these associations. The best described function of SLs in plant–microbe interactions is in the rhizosphere, where, after being exuded from the root, they activate hyphal branching and enhanced growth and energy metabolism of symbiotic arbuscular mycorrhiza fungi (AMF). Furthermore, an impact of SLs on the quantitative development of root nodule symbiosis with symbiotic nitrogen-fixing bacteria and on the success of fungal and bacterial leaf pathogens is beginning to be revealed. Thus far, the role of SLs has predominantly been studied in binary plant–microbe interactions. It can be predicted that their impact on the bacterial, fungal, and oomycetal communities (microbiomes), which thrive on roots, in the rhizosphere, and on aerial tissues, will be addressed in the near future.

Keywords *Rhizobia* · Arbuscular mycorrhiza fungi · Pathogen · Rhizosphere · Receptor · Plant hormones · Plant disease · *Medicago truncatula* · Pea · Rice · Tomato

S. Rochange
Laboratoire de Recherche en Sciences Végétales (LRSV), UMR5546, Université de Toulouse and CNRR, UPS, Castanet-Tolosan, France

S. Goormachtig
Department of Plant Biotechnology and Bioinformatics, Ghent University, Gent, Belgium

Center for Plant Systems Biology, VIB, Gent, Belgium

J. A. Lopez-Raez
Department of Soil Microbiology and Symbiotic Systems, Estación Experimental del Zaidín-Consejo Superior de Investigaciones Científicas (EEZ-CSIC), Granada, Spain

C. Gutjahr (✉)
Plant Genetics, School of Life Sciences Weihenstephan, Technical University of Munich (TUM), Freising, Germany
e-mail: caroline.gutjahr@tum.de

© Springer Nature Switzerland AG 2019   121
H. Koltai, C. Prandi (eds.), *Strigolactones - Biology and Applications*,
https://doi.org/10.1007/978-3-030-12153-2_4

4.1 The Role of Strigolactones in Arbuscular Mycorrhiza Symbiosis

Although invisible to the naked eye, the arbuscular mycorrhiza (AM) symbiosis is one of the most widespread symbiotic associations on Earth. The symbiotic partners are a group of microscopic soil fungi from the subphylum Glomeromycotina and the vast majority of land plants. A plant taken randomly in a natural or agricultural ecosystem is much more likely to be colonized by AM fungi (AMF) than not. Glomeromycotina are obligate biotrophs, meaning that they need to colonize a plant to complete their life cycle. They can penetrate the roots of host plants and colonize the root cortex while simultaneously extending their mycelium far into the soil. In root cortical cells, they develop highly branched structures called arbuscules, at which exchange of nutrients takes place (Smith and Read 2008). Thanks to their long, fine hyphae and excellent mineral uptake capacities, AMF can provide their host plants with water and mineral nutrients (especially phosphate), which would otherwise be inaccessible to the roots (Smith and Smith 2011). In return, plants provide AMF with organic carbon in the form of sugars and lipids for their growth and development (Roth and Paszkowski 2017; Keymer and Gutjahr 2018).

The development of the symbiosis requires communication and coordination between the two partners. As obligate biotrophs, AMF need to sense the presence of a host plant before they commit themselves to extensive growth and energy-consuming processes. In addition, host plants need to recognize symbiotic partners among a large range of soil microorganisms, and to adjust the extent of root colonization to their nutritional needs, in order to maintain a suitable cost/benefit ratio. Soil-diffusible compounds released by the two partners contribute to the early steps of this communication and mutual recognition. Among these, strigolactones (SLs) have been identified as important symbiotic signals of plant origin.

4.1.1 Discovery of SLs as Signals in Arbuscular Mycorrhiza Symbiosis

AM fungal spores germinate in the soil and form hyphae that can grow for only a few days in the absence of a host plant. In the vicinity of host plants, hyphae display a typical branching response and their growth continues. This phenomenon has been known for a long time and is also observed when the host plant is replaced by its root exudates, indicating the involvement of plant-released diffusible compounds (Buée et al. 2000). It took many years, however, to purify enough of these active compounds for their identification. Starting from root exudates of *Lotus japonicus*, Akiyama et al. (2005) identified the SL 5-deoxystrigol as a potent inducer of AM hyphal branching (Fig. 4.1a). After the discovery of the germination stimulant activity of SLs on seeds of root parasitic plants back in the 1960s (Cook et al. 1966), this revealed a second important function of SLs in the rhizosphere and

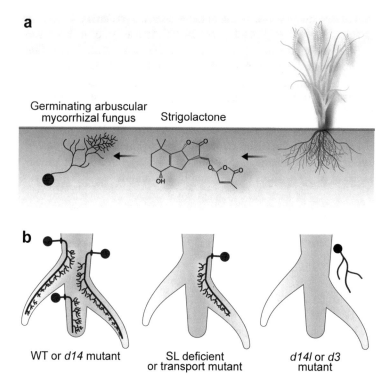

Fig. 4.1 Role of strigolactone in root AM symbiosis. (**a**) Strigolactones (SL) exuded by roots induce hyphal branching of arbuscular mycorrhiza fungi (AMF) (Akiyama et al. 2005; Besserer et al. 2006). (**b**) SL biosynthesis and transport mutants are less colonized by AMF than wild-type plants (reviewed in Waters et al. 2017). Rice *d3* and *d14l* mutants impair hyphopodium formation and are not colonized (Gutjahr et al. 2015)

provided an explanation as to why plants have persisted through evolution to secrete compounds that stimulate harmful parasites.

4.1.2 Cellular and Molecular Effects of SLs on AMF

In addition to hyphal branching, SLs also stimulate spore germination, hyphal growth, and mitosis in AMF. From a metabolic perspective, these events are concomitant with a stimulation of respiration and mitochondrial biogenesis (Besserer et al. 2006, 2008). These observations may reflect a metabolic switch of the fungus to an active state compatible with root colonization. Treatment with SLs also triggers a rapid increase in cytosolic calcium concentration in AM hyphae (Moscatiello et al. 2014). The gene expression response to SLs includes, but is not limited to, an upregulation of genes associated with respiration. Interestingly, the expression of several genes encoding putative secreted proteins is also stimulated by SLs (Tsuzuki et al. 2016; Kamel et al. 2017). At least one of the corresponding proteins, SIS1, is important for the symbiosis (Tsuzuki et al. 2016). Finally, SLs stimulate the release of short chitin oligosaccharides by AMF (Genre et al. 2013). These compounds are part of the fungal signals required for recognition of AMF by their host plants. This highlights the role of SLs as key components of the molecular dialog between the plant and fungal partners of AM symbiosis.

Some AM species host endosymbiotic bacteria. In these species, increased bacterial division and increased transcript accumulation of some bacterial genes are observed upon SL treatment (Salvioli et al. 2016), but it is not known whether these effects of SLs are direct or mediated by the fungus.

4.1.3 Strigolactone Perception by Arbuscular Mycorrhiza Fungi

The activity of SLs on AMF can be detected at extremely small concentrations (down to 10^{-13} M in some AM species) (Besserer et al. 2006). This supports the existence of a highly sensitive perception mechanism in these fungi, but to date no receptor(s) or components of the SL signaling pathway in the fungus have been identified. This identification will be a difficult task, because AMF are not amenable to classical genetics, and it is difficult to obtain large amounts of fungal biomass for biochemical approaches. Searches in AM genomic sequence data have so far failed to identify obvious homologs of the α/β-fold hydrolases known as plant SL receptors (Waters et al. 2017). If we consider the issue from a ligand perspective, the structural requirements for an SL-like compound to be active on hyphal branching are quite different from those reported for shoot branching repression activity in pea (Akiyama et al. 2010a; Boyer et al. 2012; Mori et al. 2016). The D ring seems essential for the different bioactivities of SLs (germination of parasitic seeds, hyphal branching of AMF, and repression of shoot branching in angiosperms), while the requirements on the ABC moiety differ between the three types of target organisms. This suggests some degree of divergence between the corresponding receptors but does not rule out the possibility that the AM fungal receptor(s) belong(s) to the α/β-fold hydrolase family. Alternatively, fungal and plant receptors could have arisen through convergent evolution and be structurally unrelated.

4.1.4 Importance of Strigolactones in the AM Symbiosis

SL-deficient plant mutants have been used to assess the importance of SLs in the AM symbiosis. Although the extent to which they are affected in root colonization differs between plant species, they generally display low levels of root colonization

(Fig. 4.1b), while intraradical fungal structures including arbuscules appear morphologically normal (Gomez-Roldan et al. 2008; Kohlen et al. 2012; Kretzschmar et al. 2012; Yoshida et al. 2012; Kobae et al. 2018). Therefore, SLs are necessary for a quantitatively normal level of root colonization but do not seem to be absolutely essential for the symbiosis to be established. The residual level of root colonization observed in the mutants could be attributed either to undetectable, residual amounts of SLs in these mutants or to additional active compounds present in the root exudates.

In addition to being external symbiotic signals, SLs act as phytohormones in the host plant (see Chap. 2). One might wonder whether these hormonal functions are also involved in the AM symbiosis. In contrast to SL-deficient mutants, plant mutants defective in the SL receptor D14 are not impaired in their ability to form AM. This is an indication that SLs are important in the symbiosis as rhizospheric signals targeting the fungus, rather than plant internal regulators in further stages of the symbiosis within the host.

4.1.5   Control of Strigolactone Synthesis and Exudation

Like other plant hormones, SLs are subject to a finely tuned regulation of synthesis and transport. Because SLs are also signals to soil microorganisms contributing to plant nutrition, their release must be coordinated with the host's nutritional needs. In this respect, SLs are important integrators at the crossroads of mineral nutrition, development, and plant–microbe interactions.

Consistent with the nutritional benefits of the AM symbiosis, SL biosynthesis and exudation are strongly activated under phosphate deprivation (reviewed in Carbonnel and Gutjahr 2014). This regulation occurs at a whole-plant level rather than in response to local phosphate availability. In some species, starvation in other nutrients like nitrogen also enhances SL production. Reduced SL production likely contributes to, but is not solely responsible for, the decreased rate of root colonization observed under high phosphate supply (Balzergue et al. 2010; Breuillin et al. 2010; Carbonnel and Gutjahr 2014). Reciprocally, the colonization of roots by AMF often triggers a decrease in SL production, possibly as a result of better phosphate nutrition (López-Ráez et al. 2010a, b). Alternatively, SL levels could be regulated via symbiotic signaling itself, to contribute to keeping the fungus under control and avoid excessive levels of root colonization. The reduction in SL biosynthesis could also be involved in the regulation of root architecture by AMF (Fusconi 2014; Lanfranco et al. 2018).

SL contents can also be affected by other plant hormones. The best documented example is that of gibberellins, which negatively affect SL synthesis in rice and *Lotus japonicus* (Ito et al. 2017). Indirect evidence suggests a positive effect of auxin on SL synthesis through enhanced expression of key biosynthetic genes (Foo et al. 2005). Interestingly, these effects mirror those of gibberellins and auxin on AM symbiosis (respectively, an inhibition and a stimulation of the symbiosis). Evidence for interaction between SL and abscisic acid (ABA) signaling has also been reported (reviewed in Lopez-Raez 2016) with tomato ABA biosynthesis mutants producing reduced amounts of SLs and SL biosynthesis mutants of *Arabidopsis* being hyporesponsive to ABA (López-Ráez et al. 2010a, b; Ha et al. 2014). ABA is also involved in regulating arbuscule development and quantity of root colonization (Herrera-Medina et al. 2007; Charpentier et al. 2014). It can thus be hypothesized that modulation of SL synthesis contributes to the well-known effects of these three hormones on AM interactions.

SL translocation in host plants involves PDR1, a member of the ATP-binding cassette (ABC) family of transporter proteins. PDR1 is involved in the upward transport of SLs from the root apex and in the exudation of SLs to the rhizosphere (Kretzschmar et al. 2012; Sasse et al. 2015). Similar to SL deficiency, mutations in *PDR1* decrease the level of root colonization, but do not impair qualitatively the formation of arbuscules. In Petunia, *PDR1* is expressed in hypodermal passage cells. These non-suberized cells are found in plant species with a dimorphic hypodermis, and in these species, they are the preferred route taken by AMF to reach the cortex (Sharda and Koide 2008). It has been hypothesized that the localized secretion of SLs by these cells could "guide" AM hyphae through the roots.

4.1.6   Significance of Strigolactone Diversity

There is some structural diversity among the >20 SLs identified to date (see Chap. 1), and each plant species usually produces several different SL forms. The functional relevance of this diversity has barely been explored. As far as the AM symbiosis is concerned, most tests for bioactivity have been carried out on *Gigaspora* species and based on the hyphal branching response. Interestingly, the hyphal branching pattern induced by different SLs can vary qualitatively, i.e., branches of varying order can be produced (Akiyama et al. 2010a). This suggests that several fungal signaling pathways could be recruited, but the link between hyphal branching patterns and the ability to colonize a host plant remains unclear. Future work will probably evaluate the activity of different SL forms on other biological responses in AMF, as well as on other AMF species.

To date, there is no clear indication for a role of SL diversity on the determination of host range in the AM symbiosis. In laboratory studies, this interaction does not display a high degree of host specificity. Nonetheless, AM fungal species are not functionally or ecologically equivalent. It would therefore be of interest to investigate whether the blend of SLs released by a given plant facilitates the recruitment of particular AMF species. Unfortunately, this question will remain beyond reach until appropriate tools are generated, for example, plant varieties producing tailor-made sets of SLs.

As for plant species that are nonhosts for AMF, hyphal branching inducers were long thought to be absent from their root exudates, but this assumption no longer holds true. Canonical SLs can be found in root exudates of *Lupinus albus*, for example, although their activity in inducing hyphal branching is masked by the presence of inhibitors (Kaori et al. 2008; Akiyama et al. 2010b). *Arabidopsis thaliana* produces carlactone derivatives such as carlactonoate and methyl-carlactonoate, which can induce hyphal branching in AMF (Mori et al. 2016). In any case, a reduced presence or activity of SLs does not account for the inability of such species to become mycorrhiza: symbiotic capacity cannot be restored in AM-incompetent plant species by exogenous SL application, and several key genes essential for the symbiosis are absent from their genomes (Delaux et al. 2014; Favre et al. 2014; Bravo et al. 2016).

4.1.7 Toward the Identification of Novel Symbiotic Signals

Mutants in the plant SL signaling component MAX2 (see Chap. 1) are severely affected in their mycorrhiza capacity (Yoshida et al. 2012). In contrast, root colonization levels are similar to WT, or even higher, in the SL receptor *d14* mutants (Fig. 4.1b) (Yoshida et al. 2012). These observations indicate that the roles of the SL receptor and of MAX2 in mycorrhiza symbiosis are not overlapping. This discrepancy can be explained by the fact that MAX2 is involved not only in SL perception but also in the perception of karrikins and/or unknown ligands of the KAI2 receptor (see Chap. 5). Indeed, rice plants mutated in *KAI2* are defective in colonization by AMF (Gutjahr et al. 2015). Thus, there seems to be an SL-independent but MAX2-dependent pathway essential for the AM symbiosis. The signal triggering this pathway may be of plant or fungal origin and remains to be identified. Based on the strength of the mycorrhiza phenotypes in rice, this novel signal could be even more important than SLs in the symbiosis. Its identification will be a major challenge, but the potential outcomes reach beyond the symbiosis per se, since KAI2 is also known to perceive endogenous plant signals associated with seedling development. This is a perfect illustration of how AM symbiosis and plant development are intimately intertwined.

In conclusion, it cannot be stressed enough that being colonized by mycorrhiza fungi is the normal status of most plants. The dual role of SLs as phytohormones and symbiotic signals places them in a unique position to coordinate many aspects of plant development and nutrition in AM host species. To the scientist, this dual role represents great challenges and opportunities: on one hand, it is sometimes difficult to disentangle symbiotic from hormonal effects of SLs, and on the other hand, new insight is likely to arise at the intersection of these two research fields.

4.2   The Role of Strigolactones in Root Nodule Symbiosis

Plants need reduced nitrogen compounds in the form of nitrate or ammonia to synthesize essential macromolecules. When the availability of these compounds in the soil is scarce, which often happens in natural environments, plants grow slowly. Hence, to compensate for this lack of nutrients, large amounts of chemical nitrogen are used as fertilizer in agriculture. However, this procedure imposes a pressure on the environment and is not sustainable because nitrogen fertilizers, produced by the Haber–Bosch process, require a large energy input. Additionally, by the current applications, a lot of fertilizer is wasted. Very often the nitrogen compounds are converted into nitrogen gas through microbial soil activity before the plant roots can take them up, or they leach inside the groundwater reservoir, leading to decreased quality of drinking water.

Some plant species circumvent the shortage in nitrogen compounds by establishing a mutualistic collaboration with soil bacteria. Indeed, in nature, nitrogen exists predominantly as an inert gas, N_2, that is unavailable for plants but that can be reduced to ammonia by bacteria that contain the nitrogenase enzyme in a process known as nitrogen fixation (Dos Santos et al. 2012). Most of these so-called diazotrophic bacteria or diazotrophs fix nitrogen for their own consumption, but some bacteria transfer the fixed ammonia to the plant roots, where it can be used for growth. In return, the bacteria obtain plant-derived carbon sources and a specific ecological niche inside the plant.

One such example of a nitrogen-fixing symbiosis is the interaction between legume plants and bacteria from diverse species that are collectively designated rhizobia. The outcome of this symbiosis is the formation of new root organs, the nodules, inside which the rhizobia find excellent conditions to fix nitrogen for the plant (Oldroyd and Downie 2008; Oldroyd et al. 2011). When plants sense nitrogen deprivation, they initiate molecular communication with neighboring rhizobia. To this end, legume roots secrete a particular blend of various structurally different secondary metabolites, mainly flavonoids and isoflavonoids, which function as signals to activate the so-called nodulation (Nod) genes inside the rhizobia (Broughton et al. 2000). As a result, Nod genes are induced and Nod proteins produced that together generate the Nod factors, short-chained lipochitooligosaccharides that are decorated with strain-specific chemical groups at one or both ends of the molecule. Subsequently, Nod factors are perceived by the plant roots, upon which the plant initiates the development of the nodule organ and allows rhizobial colonization (Gough and Cullimore 2011). Nodulation is charac-terized by strong host specificity, such that in many cases, only certain legumes can enter into a symbiotic association with specific rhizobial strains. For instance, the legume *Medicago truncatula* (barrel medic) relates with the bacterium *Sinorhizobium meliloti*, but not with, for instance, *Azorhizobium caulinodans* that is the symbiont of the tropical legume *Sesbania rostrata*. This narrow host range is determined by the specificity of the abovementioned molecular dialog: a particular rhizobial strain only activates its Nod genes when flavonoids with a particular structure are available, whereas legume plants will only trigger the nodulation program when they perceive Nod factors with a specific structure (Radutoiu et al. 2007; Fliegmann and Bono 2015; Liu and Murray 2016).

As soon as perfect match is made, the rhizobia will enter the roots, very often through the root hairs. The root hairs will curl to entrap the rhizobia. Through the invagination of the root cytoplasmic membrane, an infection thread forms and guides the bacteria toward the cortical root cells. Simultaneously, the cortical cells start to divide to form a nodule primordium, and inside these cells, the infection threads release the rhizobia in the form of symbiosomes, i.e., bacteria that are surrounded by a host-derived membrane (Oldroyd and Downie 2008). In these symbiosomes, nitrogen fixation takes place. Likewise, mature nodules, which can be determinate or indeterminate, are formed. Determinate nodules are round-shaped, such as those that develop on *Lotus japonicus*, *Phaseolus* sp. (bean), and *Glycine max* (soybean), and they are fully occupied by infected cells interspersed by some uninfected cells. Indeterminate nodules are elongated, and the meristem develops at the apical side of the nodule, which continuously provides new cells for infection.

4.2.1   Strigolactones Modulate the Quantity Nodulation

Bacterial infection and nodule organ development need to be strictly coordinated by an interplay between various plant hormones. Once the Nod factors are perceived by the Nod factor receptors, spatiotemporal patterns of gene expression occur to initiate rhizobial infection and cortical cell division. Almost all classical plant hormones have been shown to be involved in the development of a functional nodule and to be intertwined in a complex interaction network (Ferguson and Mathesius 2014). One group of important plant hormones regulating nodule organ development is cytokinins, as evidenced by mutant analysis: knockout mutants in a cytokinin receptor abolish nodulation, whereas spontaneous nodules appear on gain-of-function cytokinin receptor mutants (Murray et al. 2007; Tirichine et al. 2007).

First results suggest that SLs as well play a role in the nodulation process (Fig. 4.2). In many legume species, application of the synthetic SL *rac*-GR24 positively affects the nodule number, as for *Medicago sativa* (alfalfa), *Pisum sativum* (pea), and soybean (Fig. 4.2a) (Soto et al. 2010; Foo and Davies 2011; Rehman et al. 2018). Additionally, when SL biosynthesis genes are silenced or mutated, as shown for *Lotus japonicus*, pea, and soybean, fewer nodules are produced (Fig. 4.2b) (Foo and Davies 2011; Foo et al. 2013; Liu et al. 2013; Haq et al. 2017). These data imply that SLs play a positive role in nodulation. However, the picture might not be that clear, because not all data obtained in various legume species support this model. For instance, in contrast to what would be expected, the nodule number increased in the SL-insensitive *ramosus4* (*rms4*) mutant (Fig. 4.2b), which is affected in the ortholog of the *Arabidopsis thaliana MORE AXILLARY BRANCHES 2* (*MAX2*), which is part of the SL receptor complex (Foo et al. 2013). Furthermore, treatment with *rac*-GR24 affected the nodule number in a concentration-dependent manner in *M. truncatula*, with a stimulating effect at very low

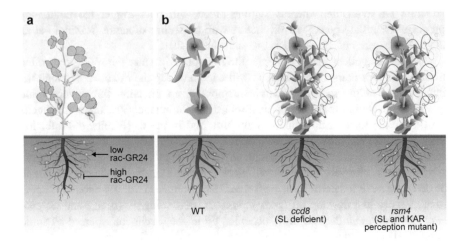

Fig. 4.2 Role of strigolactones in root nodule symbiosis. (**a**) Addition of rac-GR24 to *M. truncatula* roots increases nodule numbers at low concentrations but decreases nodule numbers at high concentrations (De Cuyper et al. 2015). (**b**) The pea strigolactone (SL) biosynthesis mutant *ccd8* carries a lower number of nodules than wild type, whereas the SL and karrikin (KAR) perception mutant *rms4* (*max2*) displays more nodules (Foo et al. 2013). Note that only one mutant allele was available precluding confirmation of the phenotype

rac-GR24 concentrations and a negative effect at higher concentrations (De Cuyper et al. 2015). Hence, these data demonstrate that SLs modulate the nodulation process. However, the SL action on the nodulation may depend on its concentration and the specific SL sensitivity of the legume species. Moreover, it must be kept in mind that MAX2/RMS4 is also involved in the perception of the smoke-derived karrikin and the unknown endogenous ligand together with KARRIKIN-INSENSITIVE 2 (KAI2) (Waters et al. 2017). Thus, it cannot be excluded that other carotenoid-derived molecules perceived by a complex of KAI2 and MAX2/RMS4 might play a role during nodulation. Furthermore, *rac*-GR24 consists of two diastereoisomers, of which two seem to be recognized by KAI2 in *Arabidopsis* (Scaffidi et al. 2014; Flematti et al. 2016). Hence, future research should carefully dissect the role of SL and karrikin signaling and integrate detailed metabolite studies to find strict correlations between plant mutants, metabolites, and function.

One hint for SL function in nodule formation might come from the tissue-specific expression patterns of the SL-related genes during nodulation (Liu et al. 2011; van Zeijl et al. 2015; Haq et al. 2017; McAdam et al. 2017). Several SL biosynthesis genes have been found to be upregulated during nodulation in an expression pattern that is controlled by the Nod factor signaling pathway. Tissue-specific analysis revealed that the promoters were active in nodule primordia, nodule meristems, and the early infection zone of mature indeterminate *M. truncatula* nodules. Because the activity occurred at the cell division sites, this expression pattern would fit with a role in the control of cell division in interplay with auxin and cytokinin. However, no changes in nodule structure have been observed in various SL mutants in different legume species, indicating that such a role might not be of major importance. This expression pattern could also be symptomatic of a role in autoregulation of nodulation (AON), the process used by legume plants to control the number of nodules per plant regulating the amount of fixed nitrogen. Nevertheless, studies with AON mutants could so far not confirm this hypothesis (Foo et al. 2014; De Cuyper et al. 2015).

SLs may already act at earlier stages of the root nodule symbiosis establishment. In *M. truncatula*, addition of rac-GR24 had a negative effect on the infection thread development and on the expression of an infection-related marker gene, whereas this negative effect was abolished in an ethylene perception mutant that causes massive infection (Breakspear et al. 2014; De Cuyper et al. 2015). On the contrary, a SL biosynthesis mutant of pea had fewer infection threads than the wild type (McAdam et al. 2017). Therefore, these data would suggest a role for SLs in the control of infection thread development. However, the contrasting results from the *rac*-GR24 treatment and the mutant analysis indicate either that the karrikin receptor complex is involved in nodulation or that a tight control of SL concentrations is required for the desired outcome of the legume–rhizobia interaction. It might also be that different legumes display a different sensitivity to SLs. In addition, the sites of hormone biosynthesis and action may not necessarily be the same. Hence, SLs might be produced in the dividing cells but act at the rhizobial infection level. A detailed cell-specific expression analysis of the genes involved in SL perception might give better insights into the validity of this hypothesis. In general, more research is needed to understand the role of SLs in nodulation.

4.2.2   Strigolactones May Additionally Stimulate Rhizobia

In addition to an endogenous function, SLs might also act as rhizosphere signals, i.e., not inside the plant cells but rather after their secretion into the soil environment. Certainly, SLs do not play such an important rhizosphere role in nodulation as they do in AM formation, because treatment with *rac*-GR24 had no effect on the bacterial growth or the Nod factor production (Soto et al. 2010). However, SLs, being present in the rhizosphere, might still affect the rhizobia, the outcome of which is only important under ecologically relevant conditions and, hence, difficult to observe in the laboratory. For instance, the use of the active SL analog 2′-*epi*-GR24 revealed that SLs stimulated the swarming motility of *Rhizobium leguminosarum*, i.e., the rapid coordinated movement of bacteria across a surface, through the activation of the flagellin (*flaA*) gene (Peláez-Vico et al. 2016). However, whether this effect causes the observed changes in nodule numbers needs further investigation.

Fig. 4.3 Example for a role of strigolactones in plant–pathogen interactions. A tomato strigolactone biosynthesis mutant (*ccd8*) is more susceptible to the necrotrophic fungus *Alternaria alternata* than the wild type (Torres-Vera et al. 2014). Note that only one mutant allele was available precluding confirmation of the phenotype. Image credit: Rocio Torres-Vera

4.3   The Role of Strigolactones in Plant–Pathogen Interactions

In addition to their role in beneficial interactions between plants and microbes, SLs have also been proposed to influence interactions with pathogenic fungi and bacteria. However, information about this topic is still scarce and sometimes controversial (López-Ráez et al. 2017). In this section, we will try to summarize what is known so far and try to convey a general idea about the potential role of SLs in pathogenic plant–microbe interactions.

4.3.1   Pathogenic Fungi

The first indication of the involvement of SLs in defense responses against pathogens appeared in 2014 (Torres-Vera et al. 2014). It was shown that leaves of the SL-deficient tomato line *Slccd8* were more susceptible than those of the corresponding wild-type plant to the airborne necrotrophic fungus *Botrytis cinerea*, a pathogen causing gray mold on a number of important agronomic crops (Fig. 4.3). *Slccd8* was also more susceptible to *Alternaria alternata*, another necrotrophic fungus which causes leaf spot disease on many plant species (Torres-Vera et al. 2014). Interestingly, *ccd7* and *ccd8* knockout mutants of the moss *Physcomitrella patens* were also more susceptible to the necrotrophic pathogenic fungus *Sclerotinia sclerotiorum* (Decker et al. 2017), responsible of the white mold disease. However, it currently appears that *P. patens* can only produce carlactone, the precursor of SLs (Yoneyama et al. 2018). Therefore, it is possible that resistance to *S. sclerotiorum* in *P. patens* is mediated by carlactone and not by SLs.

Conversely, in root–pathogen interactions, no effect of SLs on resistance to fungal pathogens has been reported. No differences in disease development were observed between the SL-deficient pea mutant *ramosus1* (*ccd8*) and its corresponding wild type after root infection with the soilborne hemibiotrophic fungus *Fusarium oxysporum* (Foo et al. 2016), which causes vascular wilt. Similarly no differences were observed for the pathogenic oomycete *Pythium irregulare*, another soilborne pathogen causing damping off and root rot in many plant species, in SL-deficient and SL-insensitive pea mutants (Blake et al. 2016). Therefore, as for other phytohormones, the role of SLs in defense responses against pathogenic fungi depends on the pathosystem examined. It seems that SLs exert their defensive role against fungal pathogens with a necrotrophic lifestyle, or alternatively, that the involvement of SL differs among plant organs (leaf vs. root). More studies must be carried out to examine these hypotheses. In addition to the plant-pathosystem and the plant organ or tissue, the experimental conditions used for the assays need to be taken into account and carefully monitored in experiments, as they can greatly affect the results.

Given the role of SLs in regulating metabolism and development of AMF, it is important to ascertain whether they exclusively influence constitutive plant defense or the plant immune response or if they also act directly on the growth of pathogenic fungi. Direct effects of SLs on hyphal growth and/or branching have been reported for a number of air- and soilborne pathogenic fungi and oomycetes when grown in vitro (Steinkellner et al. 2007; Dor et al. 2011; Torres-Vera et al. 2014; Blake et al. 2016; Foo et al. 2016; Belmondo et al. 2017). However, there are inconsistencies between the different reports. For instance, *rac*-GR24 application was reported to promote hyphal branching and to inhibit the radial growth in *B. cinerea* (Dor et al. 2011; Belmondo et al. 2017), while no effect was observed in other studies (Steinkellner et al. 2007; Torres-Vera et al. 2014). Divergent responses were also reported for *A. alternata*, showing a negative or no effect of *rac*-GR24 in fungal growth (Dor et al. 2011; Torres-Vera et al. 2014). An inhibitory effect on fungal growth of *rac*-GR24 and other synthetic SL mimics was also observed for *S. sclerotiorum*, although this was accompanied by an intense hyphal branching, similar to other fungal plant pathogens (Dor et al. 2011; Oancea et al. 2017). Different concentration ranges of *rac*-GR24 and in vitro growth conditions might explain the variability observed among the studies described above. Furthermore, as for other studies, the use of *rac*-GR24 is problematic because it contains two diastereoisomers, which based on genetic analyses seem to target to different receptors in *Arabidopsis*: one seems to target the SL receptor D14 and the other one the karrikin receptor KAI2 (Scaffidi et al. 2014). It is possible that also fungi possess receptors for both diastereoisomers and it is unclear if the observed fungal responses are true SL responses or whether they are caused by the diastereoisomer, which targets KAI2 in *Arabidopsis* (Scaffidi et al. 2014). It would be interesting to see whether natural SLs used at physiological amounts are also able of affecting the development of these fungal pathogens.

4.3.2 Pathogenic Bacteria

Even less information than for plant interactions with pathogenic fungi exists about the potential effect of SLs in plant interactions with pathogenic bacteria (López-Ráez et al. 2017). A recent study analyzed the involvement of SLs in the leafy gall syndrome in *Arabidopsis* (Stes et al. 2015), a disease caused by the biotrophic actinomycete *Rhodococcus fascians*. This syndrome affects a wide range of plant species and is characterized by delayed senescence, loss of apical dominance, and activation of dormant axillary meristems, leading to a stunted and bushy plant appearance (Stes et al. 2013). Both SL biosynthesis (*max1*, *max3*, and *max4*) and signaling (*max2*) mutants were hypersensitive to *Rhodococcus fascians* (Stes et al. 2015). The same effect was observed when a SL biosynthesis inhibitor (D2) was used. Interestingly, exogenous *rac*-GR24 application reduced leafy gall formation, suggesting that the efficiency of syndrome development depends on SL levels (Stes et al. 2015). The *Arabidopsis* signaling mutant *max2* was also reported to be more susceptible to the airborne necrotrophic bacterium *Pectobacterium carotovorum* and the hemibiotroph *Pseudomonas syringae* (Piisilä et al. 2015). These bacteria cause disease in a wide host range, including many agriculturally important plant species. They enter the host via stomata; thus stomatal closure is an important mechanism of plant defense. Interestingly, *P. syringae*-induced stomatal closure was disrupted in the *max2* mutant (Piisilä et al. 2015). *max2* has been shown to be hyposensitive to abscisic acid (ABA), displaying more open stomata in the absence of the pathogen (Ha et al. 2014). The increased susceptibility correlated with an accumulation of ABA and salicylic acid (SA) and increased sensitivity to reactive oxygen species (Piisilä et al. 2015). Therefore, the open-stomata phenotype could be a reflection of the disease progression or the reduced ABA sensitivity of the mutant. Since the SL signaling element MAX2 also operates in a SL-independent signaling pathway in with the karrikin receptor KAI2 (Smith and Li 2014) and the *kai2* mutant has been shown to be hyposensitive to ABA for stomatal closure (Li et al. 2017), it is possible that the hypersusceptibility of *max2* to *P. syringae* is caused by a defect in karrikin signaling. To assign stomatal defense against *P. syringae* to the correct signaling pathway, a careful comparison of *P. syringae* infection of *kai2*, with the SL-specific receptor mutant *d14* and with SL biosynthesis mutants or specific SL inhibitors, needs to be carried out.

4.3.3   Cross Talk of Strigolactone with Other Plant Defense Hormones

Most studies described above suggest that SLs, instead of having a direct effect on pathogen growth or disease development, might exert an indirect role through cross talk with other defense-related phytohormones. Actually, there are a number of reports indicating that SLs interact with other plant hormones, although these seem to depend on the developmental process, tissue type, and environmental/experimental conditions (Al-Babili and Bouwmeester 2015; López-Ráez et al. 2017). For instance, reduced levels of the hormones jasmonic acid (JA), SA, and ABA were found in leaves of the SL-deficient tomato line *Slccd8*, and this was correlated with an increased susceptibility to *B. cinerea* and *A. alternata* (Torres-Vera et al. 2014). In addition, an in silico analysis suggested that the expression of different SL biosynthesis genes in *Arabidopsis* and rice is regulated by these defense hormones (Marzec and Muszynska 2015). As mentioned, a relationship between MAX2 and ABA has been proposed in the interaction between *Arabidopsis* and *P. syringae* (Piisilä et al. 2015). *max2* has been shown to be less drought tolerant than the corresponding wild type, probably due to a decreased stomatal sensitivity to ABA (Ha et al. 2014; Visentin et al. 2016). Since the same effect has been observed for SL biosynthesis mutants (Ha et al. 2014) and for the karrikin receptor *kai2* (Li et al. 2017), it seems that both SL- and KAI2-mediated signaling may cross talk with ABA signaling. It still needs to be investigated whether MAX2 plays a direct role in pathogen responses or an indirect one, although the increased susceptibility of the *max2* mutant to *P. syringae* suggests an involvement in stomatal function rather than in defense responses per se.

Future studies should investigate SL cross talk with defense phytohormones and/or their signaling pathways in order to better understand how plants cope with biotic stresses, with the aim to design new and more effective control strategies against plant diseases.

4.4   Conclusion and Perspectives

The best characterized role of SLs in plant–microbe interactions is the activation of germination, growth, and hyphal branching of AMF in the rhizosphere. The release of SLs to stimulate the beneficial symbiont can also have negative effects for the plant, as SLs in the rhizosphere also activate seed germination of parasitic weeds. Breeding efforts aim at reducing SL exudation from crop plants to avoid parasitic plant infestation. It will be important to understand in how far this affects AM symbiosis negatively. Other approaches may aim at mutating changing the set of SL biosynthesis genes to synthetically switch SL types (Waters et al. 2017). Since AMF respond to a range of SL molecules (Akiyama et al. 2010a), it may be possible to design plants, which exude SL types that stimulate AMF but do not induce germination of adapted parasitic weeds.

It is still unknown, how SLs regulate nodulation and whether the effect of rac-GR24 application on infection thread formation and nodulation is caused by D14- or KAI2-mediated signaling. It will be possible to dissect this, once legume mutants in each receptor gene become available. Changing SL concentrations in roots genetically or pharmacologically resulted in partially contrasting outcomes in *M. truncatula* and pea. It is possible that the two species have a different sensitivity to SLs. Alternatively, the growth conditions could be responsible for the variation, as, for example, light intensity or nutrient concentration influences SL biosynthesis (Yoneyama et al. 2012; Nagata et al. 2015) and also the biosynthesis and signaling pathways of other hormones. Therefore, the variations in SL effects may be caused by differences in balance and/or interaction of several hormones. Systematic side-by-side comparisons of SL effects on nodulation under different growth conditions can address this hypothesis.

A role of SLs in plant–pathogen interactions is just emerging, and it still remains to be shown how far SLs affect the plant and/or the invading microbe. It will also be important to determine whether SLs improve pathogen resistance directly or via influencing physical barriers (e.g., the leaf cuticle), plant reactions to the environment (e.g., stomata closure), or the action of other plant hormones.

So far, the role of SLs has mainly been studied in binary plant–microbe associations. However, plant interactions with myriads of microorganisms at the same time and the plant genotype can influence the composition of this microbiome. Since SLs are exuded to the rhizosphere, their molecular composition and amount could have an impact on the composition of the rhizosphere community of microbiota. A first hint for a connection between SL exudation and microbiome structure was provided by Schlemper et al. (2017), who investigated the bacterial microbiome composition in a range of *Sorghum* cultivars grown in different soils. In one soil the rhizosphere microbiome of the cultivar SRN-39 was very different from all other cultivars. SRN-39 exuded large amounts of orobanchol into the rhizosphere, whereas the other cultivars exuded mainly 5-deoxistrigol and sorgomol at relatively lower amounts (Schlemper et al. 2017). Although a causal relationship between the bacterial community composition and the SL exudate quantity and type composition was not established, it is possible that orobanchol favored a different set of bacteria than 5-deoxistrigol and sorgomol. It will also be interesting to learn, whether SLs influence the community of the leaf microbiota. This may happen through their interaction with other hormones and/or their influence on plant physiology or cuticle properties.

Glossary

ABC transporters Members of a transmembrane transporter family. They often consist of multiple subunits comprising transmembrane domains and membrane-bound ATPases. Hydrolysis of ATP by the ATPases fuels energy-dependent translocation of substrates across membranes.

Actinomycete Diverse order of Gram-positive, anaerobic bacteria, which have a mycelium-like, filamentous, and branching growth habit. Some species form root nodule symbiosis with plants of the Fagales, Rosales, and Cucurbitales.

Arbuscules Tree-shaped hyphal structure, formed by arbuscular mycorrhiza fungi in root cortex cells. These structures release mineral nutrients to apoplast between arbuscule and host cell and take up lipids, delivered by the host.

Arbuscular mycorrhiza Ancient symbiosis between most land plants and fungi of the Glomeromycotina. Endomycorrhiza, in which the fungus penetrates root cortex cells to form tree-shaped arbuscules. The fungus improves plant mineral nutrition and receives lipids and carbohydrates stemming from photosynthesis in return.

Biotroph Parasite or symbiont, which colonizes a living host cell and exploits the living cell for. Example for nutrients.

Chitin N-acetyl-glucosamine polymer, which is the main component of fungal cell walls.

Flavonoids Family of chemical compounds with several phenyl-rings often containing a keto group. They are widespread in the plant kingdom and act, for example, as flower colors, as toxic deterrents of pathogens, or as attractants of rhizobia in the rhizosphere.

Haber–Bosch process An industrial process producing ammonium from molecular nitrogen and hydrogen. The process requires a catalyst (e.g., iron) and high temperature and pressure (400–500 °C; 15–25 MPa). It is named after its inventors Fritz Haber and Carl Bosch.

Hemibiotroph Plant pathogen, which first colonizes the plant in a biotrophic manner and then turns into a necrotroph.

Hyphae Thread-like structures, which form the body of fungi.

Microbiome The term microbiome describes the community of microbes colonizing certain niche including bacteria, archaea, protists, fungi, and viruses or their collective genomes.

Mutualism Interaction between a minimum of two organisms, in which both organisms profit from the collaboration.

Nectrotroph Parasite, which kills the cell of the host and feeds on the dead material.

Nodule primordia Root nodule in its earliest recognizable stage, from when cell division has started to a visible small white nodule, before the nodule is mature and functional.

Oomycetes Oomycota is a group of filamentous protist with c. 500 species. The name derives from their oversized oogonia, which contain the female gametes.

Parasitism Relationship between at least two species, in which one of the two lives on the cost of its host and causes harm to it.

Spore Unit of asexual reproduction of fungi used for dispersal and survival (e.g. for plant-interacting fungi through winter, when hosts are unavailable). Spores are an integral part of the fungal life cycle.

Rhizosphere Narrow region of the soil, which is directly attached to the root and influenced by root exudates and sloughed-off plant cells. The rhizosphere hosts a specific set of microbes, which are influenced by the root activity.

Root nodule symbiosis Symbiosis between plants of most legumes and bacteria belonging to the rhizobia or less frequent members of the Fagales, Cucurbitales, and Rosales and actinomycetes. The bacteria are hosted in membrane-surrounded compartments in cells of root nodules, which are lateral organs derived from cell division. The bacteria fix atmospheric nitrogen and provide the plant with ammonium in exchange for organic carbon.

References

Akiyama K, Matsuzaki K, Hayashi H (2005) Plant sesquiterpenes induce hyphal branching in arbuscular mycorrhizal fungi. Nature 435:824–827

Akiyama K, Ogasawara S, Ito S, Hayashi H (2010a) Structural requirements of strigolactones for hyphal branching in AM fungi. Plant Cell Physiol 51(7):1104–1117

Akiyama K, Tanigawa F, Kashihara T, Hayashi H (2010b) Lupin pyranoisoflavones inhibiting hyphal development in arbuscular mycorrhizal fungi. Phytochemistry 71(16):1865–1871

Al-Babili S, Bouwmeester HJ (2015) Strigolactones, a novel carotenoid-derived plant hormone. Annu Rev Plant Biol 66:161–186

Balzergue C, Puech-Pagès V, Bécard G, Rochange SF (2010) The regulation of arbuscular mycorrhizal symbiosis by phosphate in pea involves early and systemic signalling events. J Exp Bot 62(3):1049–1060

Belmondo S, Marschall R, Tudzynski P, López Ráez JA, Artuso E, Prandi C, Lanfranco L (2017) Identification of genes involved in fungal responses to strigolactones using mutants from fungal pathogens. Curr Genet 63(2):201–213

Besserer A, Puech-Pagès V, Kiefer P, Gomez-Roldan V, Jauneau A, Roy S, Portais JC, Roux C, Bécard G, Séjalon-Delmas N (2006) Strigolactones stimulate arbuscular mycorrhizal fungi by activating mitochondria. PLoS Biol 4(7):e226. https://doi.org/10.1371/journal.pbio.0040226

Besserer A, Becard G, Jauneau A, Roux C, Sejalon-Delmas N (2008) GR24, a synthetic analog of strigolactones, stimulates the mitosis and growth of the arbuscular mycorrhizal fungus *Gigaspora rosea* by boosting its energy metabolism. Plant Physiol 148(1):402–413

Blake SN, Barry KM, Gill WM, Reid JB, Foo E (2016) The role of strigolactones and ethylene in disease caused by *Pythium irregulare*. Mol Plant Pathol 17(5):680–690

Boyer F-D, de Saint Germain A, Pillot J-P, Pouvreau J-B, Chen VX, Ramos S, Stévenin A, Simier P, Delavault P, Beau J-M et al (2012) Structure-activity relationship studies of strigolactone-related molecules for branching inhibition in garden pea: molecule design for shoot branching. Plant Physiol 159(4):1524–1544

Bravo A, York T, Pumplin N, Mueller L, Harrison M (2016) Genes conserved for arbuscular mycorrhizal symbiosis identified through phylogenomics. Nat Plants 2:15208

Breakspear A, Liu C, Roy S, Stacey N, Rogers C, Trick M, Morieri G, Mysore KS, Wen J, Oldroyd GED et al (2014) The root hair "infectome" of *Medicago truncatula* uncovers changes in cell cycle genes and reveals a requirement for auxin signaling in rhizobial infection. Plant Cell 26 (12):4680–4701

Breuillin F, Schramm J, Hajirezaei M, Ahkami A, Favre P, Druege U, Hause B, Bucher M, Kretzschmar T, Bossolini E et al (2010) Phosphate systemically inhibits development of arbuscular mycorrhizal in *Petunia hybrida* and represses genes involved in mycorrhizal functioning. Plant J 64(6):1002–1017

Broughton W, Jabbouri S, Peret X (2000) Keys to symbiotic harmony. J Bacteriol 182 (20):5641–5652

Buée M, Rossignol M, Jauneau A, Ranjeva R, Bécard G (2000) The pre-symbiotic growth of arbuscular mycorrhizal fungi is induced by a branching factor partially purified from from plant root exudates. Mol Plant Microbe Interact 13(6):693–698

Carbonnel S, Gutjahr C (2014) Control of arbuscular mycorrhiza development by nutrient signals. Front Plant Sci 5:462

Charpentier M, Sun J, Wen J, Mysore KS, Oldroyd GED (2014) Abscisic acid promotion of arbuscular mycorrhizal colonization requires a component of the PROTEIN PHOSPHATASE 2A complex. Plant Physiol 166(4):2077–2090

Cook CE, Whichard LP, Turner B, Wall ME, Egley GH (1966) Germination of witchweed (*Striga lutea* Lour.): isolation and properties of a potent stimulant. Science 154(3753):1189–1190

De Cuyper C, Fromentin J, Yocgo RE, De Keyser A, Guillotin B, Kunert K, Boyer FD, Goormachtig S (2015) From lateral root density to nodule number, the strigolactone analogue GR24 shapes the root architecture of Medicago truncatula. J Exp Bot 66(1):137–146

Decker EL, Alder A, Hunn S, Ferguson J, Lehtonen MT, Scheler B, Kerres KL, Wiedemann G, Safavi-Rizi V, Nordzieke S et al (2017) Strigolactone biosynthesis is evolutionarily conserved, regulated by phosphate starvation and contributes to resistance against phytopathogenic fungi in a moss, Physcomitrella patens. New Phytol 216(2):455–468

Delaux P-M, Varala K, Edger PP, Coruzzi GM, Pires JC, Ané J-M (2014) Comparative phylogenomics uncovers the impact of symbiotic associations on host genome evolution. PLoS Genet 10(7):e1004487

Dor E, Joel DM, Kapulnik Y, Koltai H, Hershenhorn J (2011) The synthetic strigolactone GR24 influences the growth pattern of phytopathogenic fungi. Planta 234(2):419–427

Dos Santos PC, Fang Z, Mason SW, Setubal JC, Dixon R (2012) Distribution of nitrogen fixation and nitrogenase-like sequences amongst microbial genomes. BMC Genomics 13(1):162

Favre P, Bapaume L, Bossolini E, Delorenzi M, Falquet L, Reinhardt D (2014) A novel bioinformatics pipeline to discover genes related to arbuscular mycorrhizal symbiosis based on their evolutionary conservation pattern among higher plants. BMC Plant Biol 14(1):333

Ferguson BJ, Mathesius U (2014) Phytohormone regulation of legume-rhizobia interactions. J Chem Ecol 40(7):770–790

Flematti GR, Scaffidi A, Waters MT, Smith SM (2016) Stereospecificity in strigolactone biosynthesis and perception. Planta 243(6):1361–1373

Fliegmann J, Bono JJ (2015) Lipo-chitooligosaccharidic nodulation factors and their perception by plant receptors. Glycoconj J 32(7):455–464

Foo E, Davies NW (2011) Strigolactones promote nodulation in pea. Planta 234:1073–1081

Foo E, Bullier E, Goussot M, Foucher F, Rameau C, Beveridge CA (2005) The branching gene RAMOSUS1 mediates interactions among two novel signals and auxin in pea. Plant Cell 17 (2):464–474

Foo E, Yoneyama K, Hugill CJ, Quittenden LJ, Reid JB (2013) Strigolactones and the regulation of pea symbioses in response to nitrate and phosphate deficiency. Mol Plant 6:76–87

Foo E, Ferguson BJ, Reid JB (2014) The potential roles of strigolactones and brassinosteroids in the autoregulation of nodulation pathway. Ann Bot 113(6):1037–1045

Foo E, Blake SN, Fisher BJ, Smith JA, Reid JB (2016) The role of strigolactones during plant interactions with the pathogenic fungus *Fusarium oxysporum*. Planta 243(6):1387–1396

Fusconi A (2014) Regulation of root morphogenesis in arbuscular mycorrhizae: what role do fungal exudates, phosphate, sugars and hormones play in lateral root formation? Ann Bot 113(1):19–33

Genre A, Chabaud M, Balzergue C, Puech-Pagès V, Novero M, Rey T, Fournier J, Rochange S, Bécard G, Bonfante P et al (2013) Short-chain chitin oligomers from arbuscular mycorrhizal fungi trigger nuclear Ca^{2+} spiking in *Medicago truncatula* roots and their production is enhanced by strigolactone. New Phytol 198(1):190–202

Gomez-Roldan V, Fermas S, Brewer PB, Puech-Pages V, Dun EA, Pillot J-P, Letisse F, Matusova R, Danoun S, Portais J-C et al (2008) Strigolactone inhibition of shoot branching. Nature 455(7210):189–194

Gough C, Cullimore J (2011) Lipo-chitooligosaccharide signaling in endosymbiotic plant-microbe interactions. Mol Plant Microbe Interact 24(8):867–878

Gutjahr C, Gobbato E, Choi J, Riemann M, Johnston MG, Summers W, Carbonnel S, Mansfield C, Yang S-Y, Nadal M et al (2015) Rice perception of symbiotic arbuscular mycorrhizal fungi requires the karrikin receptor complex. Science 350(6267):1521–1524

Ha CV, Leyva-Gonzalez MA, Osakabe Y, Tran UT, Nishiyama R, Watanabe Y, Tanaka M, Seki M, Yamaguchi S, Dong NV et al (2014) Positive regulatory role of strigolactone in plant responses to drought and salt stress. Proc Natl Acad Sci U S A 111(2):851–856

Haq BU, Ahmad MZ, Ur Rehman N, Wang J, Li P, Li D, Zhao J (2017) Functional characterization of soybean strigolactone biosynthesis and signaling genes in Arabidopsis *max* mutants and *GmMAX3* in soybean nodulation. BMC Plant Biol 17:259

Herrera-Medina M, Steinkellner S, Vierheilig H, Ocampo Bote J, García Garrido J (2007) Abscisic acid determines arbuscule development and functionality in the tomato arbuscular mycorrhiza. New Phytol 175:554–564. https://doi.org/10.1111/j.1469-8137.2007.02107.x

Ito S, Yamagami D, Umehara M, Hanada A, Yoshida S, Sasaki Y, Yajima S, Kyozuka J, Ueguchi-Tanaka M, Matsuoka M et al (2017) Regulation of strigolactone biosynthesis by gibberellin signaling. Plant Physiol 174(2):1250–1259

Kamel L, Tang N, Malbreil M, San Clemente H, Le Marquer M, Roux C, Frei Dit Frey N (2017) The comparison of expressed candidate secreted proteins from two arbuscular mycorrhizal fungi unravels common and specific molecular tools to invade different host plants. Front Plant Sci 8:124

Kaori Y, Xiaonan X, Hitoshi S, Yasutomo T, Shin O, Kohki A, Hideo H, Koichi Y (2008) Strigolactones, host recognition signals for root parasitic plants and arbuscular mycorrhizal fungi, from Fabaceae plants. New Phytol 179(2):484–494

Keymer A, Gutjahr C (2018) Cross-kingdom lipid transfer in arbuscular mycorrhizal symbiosis and beyond. Curr Opin Plant Biol 44:137–144. https://doi.org/10.1016/j.pbi.2018.1004.1005

Kobae Y, Kameoka H, Sugimura Y, Saito K, Ohtomo R, Fujiwara T, Kyozuka J (2018) Strigolactone biosynthesis genes of rice are required for the punctual entry of arbuscular mycorrhizal fungi into the roots. Plant Cell Physiol 59(3):544–553

Kohlen W, Charnikhova T, Lammers M, Pollina T, Tóth P, Haider I, Pozo MJ, de Maagd RA, Ruyter-Spira C, Bouwmeester HJ et al (2012) The tomato CAROTENOID CLEAVAGE DIOXYGENASE8 (SlCCD8) regulates rhizosphere signaling, plant architecture and affects reproductive development through strigolactone biosynthesis. New Phytol 196(2):535–547

Kretzschmar T, Kohlen W, Sasse J, Borghi L, Schlegel M, Bachelier JB, Reinhardt D, Bours R, Bouwmeester HJ, Martinoia E (2012) A petunia ABC protein controls strigolactone-dependent symbiotic signalling and branching. Nature 483(7389):341–344

Lanfranco L, Fiorilli V, Venice F, Bonfante P (2018) Strigolactones cross the kingdoms: plants, fungi, and bacteria in the arbuscular mycorrhizal symbiosis. J Exp Bot 69(9):2175–2188

Li W, Kien Huu N, Ha Duc C, Chien Van H, Watanabe Y, Osakabe Y, Leyva-Gonzalez MA, Sato M, Toyooka K, Voges L et al (2017) The karrikin receptor KAI2 promotes drought resistance in Arabidopsis thaliana. PLoS Genet 13(11):e1007076

Liu C, Murray J (2016) The role of flavonoids in nodulation host-range specificity: an update. Plants 5:33

Liu W, Kohlen W, Lillo A (2011) Strigolactone biosynthesis in Medicago truncatula and rice requires the symbiotic GRAS-type transcription factors NSP1 and NSP2. Plant Cell 23:3853–3865

Liu J, Novero M, Charnikhova T, Ferrandino A, Schubert A, Ruyter-Spira C, Bonfante P, Lovisolo C, Bouwmeester HJ, Cardinale F (2013) CAROTENOID CLEAVAGE DIOXYGENASE 7 modulates plant growth, reproduction, senescence, and determinate nodulation in the model legume *Lotus japonicus*. J Exp Bot 64(7):1967–1981

Lopez-Raez J (2016) How drought and salinity affect arbuscular mycorrhizal symbiosis and strigolactone biosynthesis? Planta 243:1375–1385

López-Ráez J, Charnikhovab T, Fernández I, Bouwmeester H, Pozo M (2010a) Arbuscular mycorrhizal symbiosis decreases strigolactone production in tomato. J Plant Physiol 168:294–297

López-Ráez J, Kohlen W, Charnikhova T, Mulder P, Undas AK, Sergeant MJ, Verstappen F, Bugg TD, Thompson AJ, Ruyter-Spira C, Bouwmeester H (2010b) Does abscisic acid affect strigolactone biosynthesis? New Phytol 187(2):343–354

López-Ráez JA, Shirasu K, Foo E (2017) Strigolactones in plant interactions with beneficial and detrimental organisms: the yin and yang. Trends Plant Sci 22(6):527–537

Marzec M, Muszynska A (2015) *In silico* analysis of the genes encoding proteins that are involved in the biosynthesis of the RMS/MAX/D pathway revealed new roles of strigolactones in plants. Int J Mol Sci 16(4):6757–6782

McAdam EL, Hugill C, Fort S, Samain E, Cottaz S, Davies NW, Reid JB, Foo E (2017) Determining the site of action of strigolactones during nodulation. Plant Physiol 175 (1):529–542

Mori N, Nishiuma K, Sugiyama T, Hayashi H, Akiyama K (2016) Carlactone-type strigolactones and their synthetic analogues as inducers of hyphal branching in arbuscular mycorrhizal fungi. Phytochemistry 130:90–98

Moscatiello R, Sello S, Novero M, Negro A, Bonfante P, Navazio L (2014) The intracellular delivery of TAT-aequorin reveals calcium-mediated sensing of environmental and symbiotic signals by the arbuscular mycorrhizal fungus Gigaspora margarita. New Phytol 203 (3):1012–1020

Murray JD, Karas BJ, Sato S, Tabata S, Amyot L, Szczyglowski K (2007) A cytokinin perception mutant colonized by *Rhizobium* in the absence of nodule organogenesis. Science 315 (5808):101–104

Nagata M, Yamamoto N, Shigeyama T, Terasawa Y, Anai T, Sakai T, Inada S, Arima S, Hashiguchi M, Akashi R et al (2015) Red/far red light controls arbuscular mycorrhizal colonization via jasmonic acid and strigolactone signaling. Plant Cell Physiol 56(11):2100–2109

Oancea F, Georgescu E, Matusova R, Georgescu F, Nicolescu A, Raut I, Jecu ML, Vladulescu MC, Vladulescu L, Deleanu C (2017) New strigolactone mimics as exogenous signals for rhizosphere organisms. Molecules 22(6):1–15

Oldroyd GED, Downie JA (2008) Coordinating nodule morphogenesis with rhizobial infection in legumes. Annu Rev Plant Biol 59(1):519–546

GED O, Murray JD, Poole PS, Downie JA (2011) The rules of engagement in the legume-rhizobial symbiosis. Annu Rev Genet 45:119–144

Peláez-Vico MA, Bernabéu-Roda L, Kohlen W, Soto MJ, López-Ráez JA (2016) Strigolactones in the Rhizobium-legume symbiosis: stimulatory effect on bacterial surface motility and down-regulation of their levels in nodulated plants. Plant Sci 245:119–127

Piisilä M, Keceli MA, Brader G, Jakobson L, Jöesaar I, Sipari N, Kollist H, Palva ET, Kariola T (2015) The F-box protein MAX2 contributes to resistance to bacterial phytopathogens in *Arabidopsis thaliana*. BMC Plant Biol 15(1):53

Radutoiu S, Madsen L, Madsen E, Jurkiewicz A, Fukai E, Quistgaard E, Albrektsen A, James E, Thirup S, Stougaard J (2007) LysM domains mediate lipochitin-oligosaccharide recognition and Nfr genes extend the symbiotic host range. EMBO J 26:3923–3935

Rehman NU, Ali M, Ahmad MZ, Liang G, Zhao J (2018) Strigolactones promote rhizobia interaction and increase nodulation in soybean (Glycine max). Microb Pathog 114:420–430

Roth R, Paszkowski U (2017) Plant carbon nourishment of arbuscular mycorrhizal fungi. Curr Opin Plant Biol 39:50–56

Salvioli A, Ghignone S, Novero M, Navazio L, Venice F, Bagnaresi P, Bonfante P (2016) Symbiosis with an endobacterium increases the fitness of a mycorrhizal fungus, raising its bioenergetic potential. ISME J 10(1):130–144

Sasse J, Simon S, Gübeli C, Liu G-W, Cheng X, Friml J, Bouwmeester H, Martinoia E, Borghi L (2015) Asymmetric localizations of the ABC transporter PaPDR1 trace paths of directional strigolactone transport. Curr Biol 25(5):647–655

Scaffidi A, Waters MT, Sun YK, Skelton BW, Dixon KW, Ghisalberti EL, Flematti GR, Smith SM (2014) Strigolactone hormones and their stereoisomers signal through two related receptor proteins to induce different physiological responses in Arabidopsis. Plant Physiol 165 (3):1221–1232

Schlemper TR, Leite MFA, Lucheta AR, Shimels M, Bouwmeester HJ, van Veen JA, Kuramae EE (2017) Rhizobacterial community structure differences among sorghum cultivars in different growth stages and soils. FEMS Microbiol Ecol 93(8):1–11

Sharda JN, Koide RT (2008) Can hypodermal passage cell distribution limit root penetration by mycorrhizal fungi? New Phytol 180(3):696–701

Smith SM, Li J (2014) Signalling and responses to strigolactones and karrikins. Curr Opin Plant Biol 21:23–29

Smith S, Read D (2008) Mycorrhizal symbiosis. Academic, London

Smith SE, Smith FA (2011) Roles of arbuscular mycorrhizas in plant nutrition and growth: new paradigms from cellular to ecosystem scales. Annu Rev Plant Biol 62(1):227–250

Soto MJ, Fernández-Aparicio M, Castellanos-Morales V, García-Garrido JM, Ocampo JA, Delgado MJ (2010) First indications for the involvement of strigolactones on nodule formation in alfalfa (Medicago sativa). Soil Biol Biochem 42:383–385

Steinkellner S, Lendzemo V, Langer I, Schweiger P, Khaosaad T, Toussaint JP, Vierheilig H (2007) Flavonoids and strigolactones in root exudates as signals in symbiotic and pathogenic plant-fungus interactions. Molecules 12(7):1290–1306

Stes E, Francis I, Pertry I, Dolzblasz A, Depuydt S, Vereecke D (2013) The leafy gall syndrome induced by *Rhodococcus fascians*. FEMS Microbiol Lett 342(2):187–195

Stes E, Depuydt S, De Keyser A, Matthys C, Audenaert K, Yoneyama K, Werbrouck S, Goormachtig S, Vereecke D (2015) Strigolactones as an auxiliary hormonal defence mechanism against leafy gall syndrome in *Arabidopsis thaliana*. J Exp Bot 66(16):5123–5134

Tirichine L, Sandal N, Madsen LH, Radutoiu S, Albrektsen AS, Sato S, Asamizu E, Tabata S, Stougaard J (2007) A gain-of-function mutation in a cytokinin receptor triggers spontaneous root nodule organogenesis. Science 315(5808):104–107

Torres-Vera R, García JM, Pozo MJ, López-Ráez JA (2014) Do strigolactones contribute to plant defence? Mol Plant Pathol 15(2):211–216

Tsuzuki S, Handa Y, Takeda N, Kawaguchi M (2016) Strigolactone-induced putative secreted protein 1 is required for the establishment of symbiosis by the arbuscular mycorrhizal fungus Rhizophagus irregularis. Mol Plant Microbe Interact 29(4):277–286

van Zeijl A, Liu W, Xiao TT, Kohlen W, Yang WC, Bisseling T, Geurts R (2015) The strigolactone biosynthesis gene DWARF27 is co-opted in Rhizobium symbiosis. BMC Plant Biol 15:260

Visentin I, Vitali M, Ferrero M, Zhang Y, Ruyter-Spira C, Novak O, Strnad M, Lovisolo C, Schubert A, Cardinale F (2016) Low levels of strigolactones in roots as a component of the systemic signal of drought stress in tomato. New Phytol 212(4):954–963

Waters MT, Gutjahr C, Bennett T, Nelson DC (2017) Strigolactone signaling and evolution. Annu Rev Plant Biol 68(1):291–322

Yoneyama K, Xie X, Kim HI, Kisugi T, Nomura T, Sekimoto H (2012) How do nitrogen and phosphorus deficiencies affect strigolactone production and exudation? Planta 235:1197–1207

Yoneyama K, Xie X, Yoneyama K, Kisugi T, Nomura T, Nakatani Y, Akiyama K, McErlean CSP (2018) Which are the major players, canonical or non-canonical strigolactones? J Exp Bot 69 (9):2231–2239

Yoshida S, Kameoka H, Tempo M, Akiyama K, Umehara M, Yamaguchi S, Hayashi H, Kyozuka J, Shirasu K (2012) The D3 F-box protein is a key component in host strigolactone responses essential for arbuscular mycorrhizal symbiosis. New Phytol 196(4):1208–1216

Chapter 5
Evolution of Strigolactone Biosynthesis and Signalling

Sandrine Bonhomme and Mark Waters

Abstract Studying evolutionarily primitive organisms with simpler genomes can provide information about the core genetic machinery required for any biological process, including hormone production and perception. In this chapter, we present findings on strigolactone biology based on work with two model byrophytes, the moss *Physcomitrella patens* and the liverwort *Marchantia polymorpha*. We summarise the existing knowledge of strigolactone biosynthesis in primitive plants, and discuss the role of strigolactones in regulating growth in response to competition from neighbouring plants. We then turn to strigolactone perception and signal transduction, with a focus on the diversity among putative strigolactone receptors in the KAI2/DWARF14 family of α/β-hydrolases. We speculate on the "original" role for strigolactones for early land plants as a rhizosphere signal, before they were adopted as hormones to regulate development. Finally, we summarise discoveries that explain how strigolactones released by plant roots came to be exploited as germination signals by root-parasitic weeds.

Keywords Moss · *Physcomitrella* · Liverwort · *Marchantia* · Hormone

5.1 Introduction

The origin of land plants was a defining event in the evolution of life on Earth, paving the way for animal life to emerge from the water shortly afterwards. Land plants are a monophyletic group: the successful invasion of land happened about 450–500 million years ago in a single lineage whose descendants include all extant land plants. As a result, all land plants have common genomic ancestry, which also implies shared molecular mechanisms for developmental regulation of plant growth. Since the

S. Bonhomme
Institut Jean-Pierre Bourgin, INRA, AgroParisTech, CNRS, Université Paris-Saclay, Versailles Cedex, France

M. Waters (✉)
School of Molecular Sciences, The University of Western Australia, Perth, WA, Australia
e-mail: mark.waters@uwa.edu.au

© Springer Nature Switzerland AG 2019
H. Koltai, C. Prandi (eds.), *Strigolactones - Biology and Applications*,
https://doi.org/10.1007/978-3-030-12153-2_5

emergence onto land, plants have become increasingly more complex in terms of structural diversity of organs, vascularisation, reproductive strategies, biosynthetic pathways, and so on. Accordingly, the underlying molecular pathways that bring about these changes have also become more complex. The number of transcription factor families within land plants, for example, typically increases with organismal complexity and is dramatically higher in land plants than in aquatic algal relatives.

Land plants are typically split up into eight major monophyletic groups (Fig. 5.1). The precise phylogenetic relationship between the first three, the liverworts, the

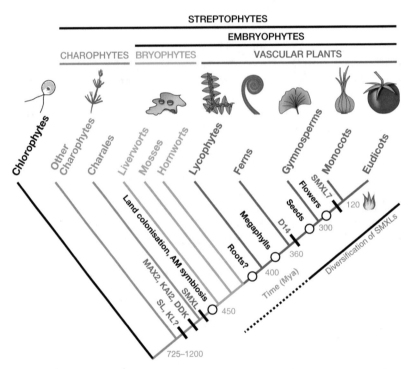

Fig. 5.1 Evolutionary timeline of land plants and strigolactone-related innovations. Approximate time points are shown in blue. Orange text indicates emergence of key SL-related innovations. SLs are produced by members of the sister group of the land plants, the Charales, but not by other charophyte algae, suggesting that SLs originated 725–1200 Mya, when the streptophytes emerged within the chlorophytes. Homologues of KAI2 and DDK in *Nitella* spp. (Charales) suggest that the SL receptor emerged at the same time as the SLs; however, the ligand specificity of such KAI2 homologues has not been established, and a KL may also have emerged at a similar time or before. Unambiguous D14 sequences emerged in the ancestor of seed plants, but other DDK proteins may have served as a SL receptor before this time (see Fig. 5.4). MAX2 emerged prior to land plants, probably subsequent to the KAI2/DDK family, but its role in SL signalling may have come much later in land plant evolution. The SMXL family emerged in the bryophytes but has undergone significant expansion and diversification over time, potentially allowing for finer control over growth and development in angiosperms, where the SMXL7/D53 clade emerged. Abbreviations: CCD8, carotenoid cleavage dioxygenase8; D14, DWARF14; DDK, D14 DLK2 and KAI2; KAI2, karrikin-insensitive2; KL, endogenous KAI2 ligand; MAX2, more axillary growth2; SL, strigolactone; SMXL, SMAX1-like. Figure reproduced and modified from Waters et al. (2017) Annual Review of Plant Biology 68, 291–322. Additional data from Bythell-Douglas et al. (2017). BMC Biol 15, 52 and Walker and Bennett (2017) BioRxiv doi:https://doi.org/10.1101/228320

hornworts, and the mosses (collectively known as the bryophytes), is contentious and unresolved, but these plants are recognised as being representatives of the oldest land plant groups. They are characterised by their small growth habit, intolerance to desiccation, no vascular tissue for conducting water, a simple rooting system known as rhizoids, and by spending most of their life cycle in the gametophytic (haploid) stage. All other plants—the tracheophytes—have vascular tissue, an innovation that took place about 400 million years ago, and a life cycle in which the diploid sporophyte is dominant. Within this group are the clubmosses (or lycophytes), the true ferns (or monilophytes), the gymnosperms, and most recently the angiosperms or flowering plants, which can be split up into monocotyledons and dicotyledons. Each of these groups is distinguished by key evolutionary innovations that have accompanied and driven an increase in diversity and complexity. Modern angiosperms dominate nearly every biome on Earth and constitute 250,000 to 400,000 species, or around 94% of all land plant species.

Plant hormones are a crucial means for plants to co-ordinate growth and development with external environmental conditions, allowing local stimuli to be translated into responses elsewhere on the plant body. Some degree of hormonal control of growth is common to all land plants. The recent analysis of the complete genome of the liverwort *Marchantia polymorpha* has shown that even early land plants had the necessary parts for producing and detecting auxin, gibberellin, cytokinin, and ethylene, among others (Bowman et al. 2017). The discovery of strigolactones as a plant hormone in 2008 soon prompted questions about the evolutionary origins of these compounds. When during plant evolution did strigolactone production occur? How did a carotenoid cleavage product become a systemic signalling molecule? Were strigolactones always a hormone, or did they originally have some other functions? And where did the receptor protein and other signalling machinery come from?

Much of what we know about strigolactone function comes from genetic studies in angiosperms such as rice (*Oryza sativa*), garden pea (*Pisum sativum*), petunia (*Petunia hybrida*), and the model plant *Arabidopsis thaliana*. Angiosperms form the basis of agriculture and constitute the dominant source of primary energy in most terrestrial ecosystems. Nevertheless, a great deal about plant development can be learned from studying simpler, more primitive plants such as mosses and liverworts and their algal relatives. Comparative genomics allows us to distinguish between conserved, "core" genetic elements that control growth of all land plants and those that are innovations of more recent plant taxa. Doing so allows us to infer the evolutionary steps that took place during the diversification of land plants. It also provides a framework for interpreting the direction of change during evolution and linking this to changes in genomic content. Both mosses and liverworts are amenable to genetic manipulation, including the production of targeted knockouts, which allows us to study gene function and test hypotheses relating to plant evolution.

In this chapter, we present the current knowledge on when and how strigolactones originated and what functions they have in non-seed plants. This field is still much in its infancy, but rapid progress in the acquisition and interpretation of genomic information is helping to unravel a number of mysteries.

5.2   Model Systems

The moss *Physcomitrella patens* (*P. patens*) has long served as a model for studies of plant development in bryophytes. A major reason for the popularity of *P. patens* is its high rate of homologous recombination, allowing gene-targeting experiments and thus making reverse genetic approaches (i.e. knockout/knock-down) relatively straightforward. The genome sequence of *P. patens* was published in 2008. *P. patens* is easily cultivated in laboratory conditions on Petri dishes, and, as the majority of its life cycle is haploid, mutant phenotypes are directly observed in the first generation following mutagenesis. *Physcomitrella* spores germinate and produce filaments called protonema, which are a single cell in thickness and elongate by tip growth in a single plane. In this way, growth at the protonemal stage is limited to two dimensions across the substrate. The youngest protonemata are rich in chloroplasts and hence are called chloronema. These filaments subsequently differentiate into caulonema, which can branch to produce new tips that grow in a new direction. The branching of caulonema and an increased growth rate are responsible for the radial expansion of the plant. After a period of radial growth, buds and then leafy shoots (gametophores) develop upon the caulonema, enabling the plant to grow vertically and, therefore, in three dimensions. Under favourable environmental conditions (e.g. reduction in temperature), male and female reproductive organs will appear at the apex of gametophores, producing motile gametes that will fuse in a diploid zygote. The sporophyte, which develops from mitotic divisions of the zygote cells, is a small capsule in which meiosis will occur, leading to spores and the next haploid generation.

More recently and along with *P. patens*, the liverwort *Marchantia polymorpha* has emerged as another model for the study of plant development in basal land plants. The *Marchantia* genome sequence was released in 2017, and therefore our functional understanding of the genome is somewhat further behind that of *Physcomitrella*. Interestingly, the number of gene copies is often lower in *Marchantia* compared to *Physcomitrella* or other mosses, which have undergone whole-genome duplication events in their evolutionary past and therefore show extensive paralogy in some gene families. Depending on the genes in question, this difference may favour *Marchantia* as a model for the systematic analysis of gene function using reverse genetics. Like *P. patens*, *Marchantia* is haploid for the majority of its life cycle, and the use of CRISPR-Cas9 technology efficiently yields targeted knock-down mutants. Although these species are not fully representative of all bryophytes, genetic analyses in *Physcomitrella* and *Marchantia* continue to yield powerful insights into the early evolutionary events relating to strigolactones in plants.

5.3 Strigolactone Biosynthesis and Functions in Primitive Plants

Like other plant hormones, strigolactones are present in very tiny amounts in plants, and therefore detection and quantification of strigolactones are challenging. Early reports of the presence of strigolactones in charophyte algae, liverworts, and mosses have been questioned as quantification methods have become more sensitive (Yoneyama et al. 2018b). As such, it is easier to infer the production of strigolactones using indirect approaches. Such evidence includes biological assays, where plant exudates or extracts are tested for the ability to induce germination of parasitic plant seeds or hyphal branching of arbuscular mycorrhizal fungi. In addition, the presence of genes encoding the key enzymes for strigolactone biosynthesis in genomes or transcriptomes provides indirect evidence for strigolactone biosynthesis. Genes encoding the enzymes for the first steps of strigolactone biosynthesis are present in all land plants. Related (ancestral) sequences have been found in algae that may encode proteins with different structures and different enzymatic roles. As relatively few complete genome sequences for algae are available, strigolactone production is still ambiguous in these clades (Table 5.1).

The first genetic evidence that bryophytes produce a strigolactone-like compound that affects plant development came in 2011 with the knockout of the *CCD8* gene in *P. patens* (Proust et al. 2011). Mutants in orthologous genes in pea, *Arabidopsis*, rice, and petunia (i.e. seed plants) show a common strigolactone-deficient phenotype typified by increased shoot branching and reduced plant height. The moss *Ppccd8* mutant, isolated through homologous recombination, showed several phenotypes, compared to its wild-type counterpart: first, the spores germinated earlier, and then the caulonema filaments expanded quicker and branched more frequently than the wild-type filaments, leading to a plant of a larger diameter (Fig. 5.2). The evidence that these phenotypes were due to a deficiency in strigolactones came from a "chemical complementation" experiment in which GR24, a synthetic strigolactone, was added to the growth medium. This treatment restored the size of *Ppccd8* mutant back to wild-type size. Moreover, genetic complementation of the mutant phenotype was obtained by expressing the pea *CCD8* sequence, under the control of a strong promoter, in the *Ppccd8* mutant. This further demonstrated that SL deficiency was responsible for the mutant phenotype and that *PpCCD8* is functionally orthologous to *CCD8* from seed plants.

Interestingly, some features observed in strigolactone-deficient mutants of seed plants were also found in the *Ppccd8* mutant. First, the *PpCCD7* gene showed evidence of transcriptional feedback, in which *CCD7* transcripts are overexpressed in SL-deficient plants relative to wild type. Second, the *Ppccd8* mutant showed more branching at the base of the gametophore compared to WT, a phenotype that is reminiscent of the increased axillary branching phenotype of *ccd8* mutants in angiosperms. Recent research in moss has indicated that this form of gametophore branching involves a complex interplay between auxin and cytokinin hormones, and a similar relationship between these two hormones also regulates shoot branching in

Table 5.1 Taxonomic distribution of key genes relating to strigolactone biosynthesis and signalling

	Angiosperms	Charophytes	Liverworts	Mosses	Lycophytes	Monilophytes	Gymnosperms	Function of encoded protein
SL signalling								
MAX2	Y	Y	Y	Y	Y	Y	Y	F-box component of the SCF complex
D14	Y	N	N	N	N	N	Y	α/β-hydrolase that binds and hydrolyses SL; serves as receptor
DDK	N[a]	N[a]	Y	Y	Y	Y	Y	α/β-hydrolases similar to D14 but function uncharacterised
eu-KAI2	Y	N[a]	Y	Y	Y	Y	Y	α/β-hydrolase that binds unknown ligand (KL); required for karrikin signalling in angiosperms
SMXL family[b]	Y	N	Y	Y	Y	Y	Y	Proteolytic targets of SL signalling; similar to class I ATPase but may be transcriptional regulator
SL biosynthesis								
D27	Y	N	Y	Y	Y	Y	Y	Carotenoid isomerase
MAX3 (CCD7)	Y	?[c]	Y	Y	Y	N?[d]	Y	Carotenoid cleavage dioxygenase
MAX4 (CCD8)	Y	?[c]	Y[e]	Y	Y	Y	Y	Carotenoid cleavage dioxygenase
MAX1	Y	Y	Y[e]	Y[f]	Y	Y	Y	Cytochrome P450 enzyme; acts upon carlactone
LBO	Y	N	Y	Y	Y	Y	Y	2-oxoglutarate- and Fe(II)-dependent dioxygenase; acts downstream of MAX1
SL production	Y	?	Y	Y	Y	Y	Y	

Table based on genomic and transcriptome data collated and analysed by Walker and Bennett (2017)

[a] Unambiguous homologues of DDK and KAI2 are present in charophytes, but the DDK/eu-KAI2 split occurred after or with emergence of land plants

[b] The SMXL7/D53 clade that has been proven to be involved in SL signalling is only present in angiosperms, but wider SMXL family members are present in all land plants

[c] CCD7- and CCD8-like sequences are present in charophytes, but they are relatively divergent compared with land plant versions

[d] Low expression levels may lead to absence from transcriptome assemblies, thus not reflecting genomic content

[e] Absent in *Marchantia polymorpha* (but present in relatives)

[f] Absent in *Physcomitrella patens* (but present in relatives)

angiosperms. Although the exact mechanisms relating to the control of auxin transport may differ between moss and angiosperms, it is noteworthy that strigolactones may affect this process in a similar manner in such diverse taxa. This is all the more surprising given that branching of gametophores in moss and shoot branching via axillary meristems in angiosperms are not homologous processes, as they occur at different stages of the life cycle (gametophyte in moss and sporophyte in angiosperms).

A striking phenotype was observed in the *Ppccd8* mutant that implied a special role for SL in communication between individuals. In wild-type *P. patens*, and in mosses generally, the density of plants determines the overall size of each plant, such that the plants on a crowded Petri dish are much smaller than those on a sparsely populated dish. Thus, moss plants can regulate their growth according to neighbouring competition. This phenomenon is reminiscent of bacterial quorum sensing, a cell-to-cell communication system based on secreted low-molecular-weight molecules that influence various cellular behaviours in response to population density. In contrast to wild-type plants, the *Ppccd8* mutant did not show an ability to control its size: plants continued to expand on the medium, such that filaments overlapped those of neighbours, regardless of their proximity (Fig. 5.2). Crucially, the presence of a wild-type moss plant could inhibit the growth of *Ppccd8* mutants towards it, suggesting that the mutants lacked the production of a signal, but not the ability to respond to it. Overall, the study of *CCD8* function in moss has revealed that strigolactones (or strigolactone-like compounds derived from carlactone) have an endogenous, developmental function in basal land plants. This finding suggests that a hormonal signalling function for strigolactones may be a very ancient feature common to all land plants and one that has been elaborated upon during plant evolution.

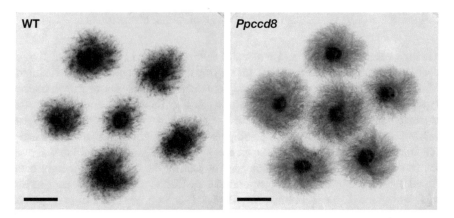

Fig. 5.2 Regulation of plant size in moss by strigolactones. Phenotype of the *Physcomitrella patens Ppccd8* mutant deficient in strigolactone synthesis (right), compared to that of wild type (left). Plants shown are 5 weeks old and grown on sterile nutrient medium. Scale bar = 1 cm

Finally, the structural diversity of strigolactones presents an additional complication for concluding whether or not a particular taxon produces these compounds. There is now a distinction between canonical strigolactones (those with a four-ring, ABC-D structure) and noncanonical strigolactones, which lack the ABC tri-cycle but retain the bioactive enol ether butenolide moiety (Yoneyama et al. 2018b). Although both types are derivatives of the biosynthetic intermediate compound carlactone (and hence carotenoids), the various steps that account for the two types of strigolactone are not fully deciphered. Functional analysis of *MAX1* homologues in non-seed plants has extended only as far back as the lycophyte *Selaginella moellendorffii*. The two MAX1 enzymes from *S. moellendorffii* can generate both canonical strigolactones (4-deoxyorobanchol) and noncanonical strigolactones (carlactonoic acid), suggesting that MAX1 performs an early step in the production of both classes (Yoneyama et al. 2018a). Furthermore, one of these MAX1 genes can complement the *Arabidopsis max1* mutant phenotype, suggesting that the function of this enzyme is highly conserved over 400 million years (Challis et al. 2013). Currently, it is unclear if this evolutionary conservation extends further back in time: although carlactone has been detected in *P. patens* (Yoneyama et al. 2018b), it is possible that canonical strigolactones are not made by this species, because *MAX1* appears to be absent from the *P. patens* genome. Nevertheless, *P. patens* itself may be unusual among mosses, given that its relatives do have *MAX1* homologues (Table 5.1). As discussed above, it is clear that some strigolactone-like, carotenoid-derived compound regulates the development of *P. patens*, even if it is generated in a MAX1-independent manner. Therefore, the biosynthetic steps and the final synthesised (signalling) molecules still need to be identified and characterised for many species, in particular the bryophytes.

5.4 Evolution of Strigolactone Signalling Mechanisms

Through extensive genetic and molecular studies, mostly in angiosperms, a good deal is now understood about how plants perceive and respond to strigolactones. Mutants defective in strigolactone perception and/or response show growth phenotypes similar to biosynthetic mutants—typically reduced overall height and increased shoot axillary growth—with the crucial distinction that they are also insensitive to exogenously supplied strigolactone. This criterion allowed the early identification of MAX2 in *Arabidopsis*, followed by D14 and then D53 in rice over the space of about a decade. Orthologues of each of these components have been found in diverse angiosperm species, consistent with a highly conserved mechanism of strigolactone response. In brief, MAX2 is an F-box protein that forms one part of the tripartite SKP1-CULLIN-F-BOX (SCF) E3 ubiquitin ligase complex. *max2* mutants of *Arabidopsis* are insensitive to strigolactones but also show several strigolactone-independent phenotypes because MAX2 is also a component of other signalling pathways (i.e. karrikins; see below). The strigolactone receptor, D14, is an α/β-fold hydrolase that was discovered by studying another strigolactone-insensitive mutant of rice called *dwarf14* (Arite et al. 2009). Orthologous mutants in *Arabidopsis* (*Atd14*), pea (*rms3*), petunia (*dad2*), and more are all fully insensitive to strigolactones. Finally, because SCF complexes are associated with targeting proteins by polyubiquitination for proteasomal degradation and several SCFs are integral parts of other plant hormone signalling mechanisms, it was predicted that strigolactone response might involve turnover of a protein(s). In 2013, another strigolactone-insensitive rice mutant called *dwarf53* was described. This mutant had the notable feature of having a dominant pattern of inheritance. Subsequent cloning of the affected gene led to the discovery of a gain-of-function insertion-deletion (indel) in the coding sequence that rendered the mutant D53 protein stable, whereas the wild-type protein is rapidly degraded in the presence of strigolactone (Jiang et al. 2013; Zhou et al. 2013). The current mechanistic model holds that D53 (and its three *Arabidopsis* homologues SMXL6, SMXL7, and SMXL8) is targeted for degradation in a MAX2- and D14-dependent manner, through ubiquitination by the SCF complex (Fig. 5.3; see also Chap. 1). It is this reduction of D53/SMXL levels that results in physiological changes in the plant, such as repression of bud outgrowth. Many of the events downstream of D53 are not yet fully characterised, but a popular hypothesis is that D53 functions as a transcriptional corepressor through associations with TOPLESS/TOPLESS-related proteins and transcription factors within the SQUAMOSA promoter-binding protein-like (SPL) family (Wang et al. 2015; Soundappan et al. 2015). It is also possible that D53 and related proteins have modes of action that do not involve the regulation of transcription.

With regard to plant evolution, when did these components appear, and what can we infer about the order of events? For example, did strigolactone biosynthesis predate perception by a receptor, or was the receptor already in place? By answering these questions, we can also draw conclusions about the importance of strigolactones in shaping the development of plant form over time.

5.4.1 Strigolactone Receptor Proteins: D14 and Its Homologues

To date, functional genetic analysis of D14 proteins has only been performed in angiosperms. D14 and its bona fide orthologues belong to a larger family of related α-/β-fold hydrolases that has members throughout the embryophytes and in algal sister groups. This family also includes a related receptor protein called karrikin-insensitive2 (KAI2). In *Arabidopsis*, *kai2* mutants show a range of developmental phenotypes including delayed seed germination and abnormal seedling and leaf development (Waters et al. 2012). In rice, KAI2 is required for establishing symbiotic interactions with arbuscular mycorrhizal fungi (Gutjahr et al. 2015). KAI2 is the likely receptor for karrikins, a collection of small butenolide compounds produced from partial combustion of plant material. Karrikins promote seed germination in many species, especially those subjected to natural fire regimes. This activity has

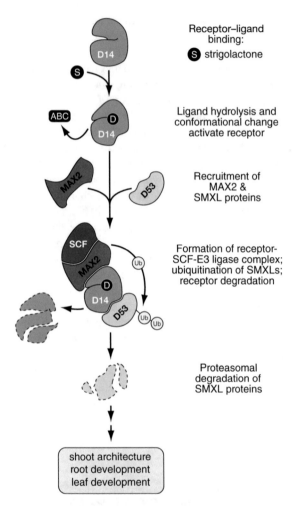

Receptor–ligand
binding:
S strigolactone

Ligand hydrolysis and
conformational change
activate receptor

Recruitment of
MAX2 &
SMXL proteins

Formation of receptor-
SCF-E3 ligase complex;
ubiquitination of SMXLs;
receptor degradation

Proteasomal
degradation of
SMXL proteins

shoot architecture
root development
leaf development

Fig. 5.3 Main events in the strigolactone perception and signalling pathway. Strigolactone or strigolactone-like compounds bind to the open form of the D14 receptor. Hydrolysis of the bound strigolactone ligand by the enzymatic activity of D14 triggers a conformational change in the protein, which presumably is necessary to recruit and stabilise an interaction with MAX2 and proteins in the SMXL7 clade (here represented by D53 from rice). MAX2 also interacts with other elements of the SCF complex, which has ubiquitin ligase activity and transfers ubiquitin monomers to target proteins. Upon formation of the SCF^{MAX2}- D14-D53 complex, D53 is degraded via the 26S proteasome. At the same time, possibly by the same ubiquitin-dependent mechanism, the D14 receptor is also degraded. Removal of SMXL7 proteins leads to physiological and developmental changes associated with the strigolactone syndrome, such as modifications in shoot architecture. All of this activity is assumed to happen in the cell nucleus based on the nuclear localisation of MAX2 and SMXL proteins. Note that the pathway depicted here describes the mechanism as understood in angiosperms but may differ in non-seed plants. Figure reproduced and modified from Waters et al. (2017). Annual Review of Plant Biology 68, 291–322

been hypothesised to result from karrikins mimicking an endogenous butenolide compound (KAI2 ligand, or KL) that all plants produce as a plant growth regulator and that is perceived by KAI2. KL is not thought to be derived from strigolactones or to have a common biosynthetic origin, as strigolactone-deficient mutants do not share *kai2* mutant phenotypes. Nevertheless, KAI2 can mediate responses to strigolactone-like compounds, suggesting that KAI2 and D14 perceive similar types of small molecule with a butenolide moiety (Scaffidi et al. 2014).

Early analyses concluded that KAI2 is present in all land plant genomes, but D14 emerged only during the evolution of seed plants, because mosses and liverworts do not contain unambiguous D14 sequences (Waters et al. 2012; Delaux et al. 2012). This led to the inference that D14 evolved from KAI2, probably as a result of a gene duplication event prior to the emergence of seed plants. Given that strigolactones are present and biologically active in mosses, this would imply that in non-seed plants, which lack D14 orthologues, KAI2-like proteins likely serve as strigolactone receptors. However, more extensive phylogenetic analysis with over 300 sequences from a wide range of taxa—comprising algae and all major land plant groups—has recently attempted to resolve the relationship between KAI2 and D14, and the picture that emerges is, predictably, a more complex and realistic one (Walker and Bennett 2017).

The identity of the closest sister group to land plants is contentious, but several orders in the charophyte algae (Klebsormidiales, Charales, Coleochaetales, and Zygnematales) are good candidates. Sequences with superficial similarity to KAI2 are present in all four of these groups, suggesting that KAI2 was present before land plants evolved. Analyses suggest that there was a very ancient split in the D14/KAI2 family, an event that occurred concomitantly with the emergence of the land plants. This split gave rise to two major clades (Fig. 5.4). The first contains the bona fide KAI2 members ("eu-KAI2") with highly conserved sequences from angiosperms, gymnosperms, monilophytes, lycophytes, mosses, and liverworts. This clade is especially invariant compared with the second clade ("DDK"), a comparatively diverse group comprised of multiple subclades that exhibit substantial sequence divergence relative to eu-KAI2 members. In contrast to the early analyses in which D14 was thought to have evolved during seed plant evolution and thus could be considered a subclade of eu-KAI2, this more extensive analysis places D14 within the DDK clade (Bythell-Douglas et al. 2017). This clade also contains a large number of D14-like (DLK) sequences, whose functions are not yet known and are relatively divergent compared with eu-KAI2 and D14 but are sufficiently widely distributed to suggest that they are not mere pseudogenes. Like the eu-KAI2 clade, the DDK clade also has members from the gamut of land plant taxa, most notably mosses, liverworts, and monilophytes. It is not yet clear what function the eu-KAI2 members and DDK members have in these basal land plant taxa, but species such as *P. patens* and *M. polymorpha* are excellent model systems for asking such questions. *Marchantia* in particular is a very promising candidate for investigation, because it contains just two sequences in this family: one is a member of the eu-KAI2 clade and the other of the DDK clade.

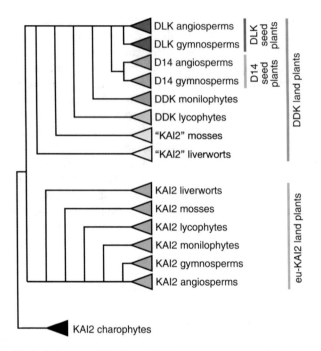

Fig. 5.4 Simplified phylogeny of KAI2 and D14 receptor proteins in land plants. Two super-clades, eu-KAI2 ("true KAI2") and DDK ("D14, DLK2 and KAI2"), diverged early in land plant evolution. All major land plant groups are represented in both clades, with the exception of hornworts, whose sequences either fit within the eu-KAI2 clade or as a sister group to all other land plant sequences (not shown). True D14 sequences are present only in seed plants and are a sister group to DLK2 and DLK3 sequences also found in seed plants (collectively termed "DLK" here). Some DDK members in mosses and liverworts have previously been assigned as KAI2 sequences and hence are labelled as "KAI2" here, even though they are not placed within the eu-KAI2 clade. While D14 proteins and some other DDK proteins are most probably genuine strigolactone receptors, the role of DLK proteins is unknown. Therefore strigolactone receptors are only a small subset of a much wider family of homologous proteins with ancient origins. Phylogeny based on data from Bythell-Douglas et al. (2017). BMC Biol 15, 52

In summary, the current picture indicates that the split in the gene family that ultimately resulted in divergence between KAI2 and D14 in derived angiosperms actually took place very early in plant evolution. Therefore, with respect to strigolactone perception, it is tempting to conclude that land plants have maintained a distinct set of strigolactone receptors, possibly predating strigolactone production via CCD8. This would also imply that eu-KAI2 proteins have long been receptors for KL, and not for strigolactones. However, as discussed below, there are much more recent and significant exceptions to this rule. Irrespective of subsequent evolutionary events, the extensive conservation of eu-KAI2 proteins raises the exciting possibility that KL is another ancient hormonal signal that is possibly at least as old as strigolactones.

5.4.2   The Function of MAX2 in Basal Land Plants

MAX2 orthologues are found in Charales and in all known genomes of embryophytes; therefore, *MAX2* probably emerged at around the same time as other components of the SL signalling pathway (Fig. 5.1). In most plant genomes, including bryophytes, *MAX2* is present as a single gene copy, but occasionally there are two copies (e.g. petunia, *Populus tremula*, *Selaginella moellendorffii*), likely as a result of lineage-specific and relatively recent whole-genome duplication events. In angiosperms such as *Arabidopsis*, MAX2 has multiple functions, both SL-related and SL-independent. In angiosperms, plants lacking MAX2 function have enhanced shoot branching phenotypes similar to *D14* mutants. However, *max2* mutants also have additional phenotypes that reveal the SL-independent functions of MAX2. Among these is the regulation of photomorphogenesis, or light-dependent development: *Arabidopsis max2* seedlings have elongated hypocotyls relative to wild-type seedlings, indicative of an impaired response to light (Nelson et al. 2011; Shen et al. 2012). SL-deficient mutants (e.g. *ccd8*) do not exhibit such defects. Therefore a crucial question regarding the evolution of plant development relates to the acquisition of the distinct functions of MAX2. Until very recently, it was not known whether MAX2 in basal land plants functions in SL perception, photomorphogenesis, or both.

Targeted gene knockout experiments were used to generate mutations in the single homologue of *MAX2* in *P. patens*. In an unexpected outcome, *Ppmax2* mutants were still able to perceive racemic GR24 (a synthetic strigolactone analogue) and did not show the expanded and branched filament phenotype of *Ppccd8* SL synthesis mutants (Lopez-Obando et al. 2018). Indeed, the *Ppmax2* mutant exhibits a phenotype that is quite the opposite, with much shorter filaments and rapidly differentiating gametophores that are larger and fewer in number relative to wild type. Crucially, the *Ppmax2* mutant gametophores are elongated in red light, a phenotype that is reminiscent of the hypocotyl of *Arabidopsis max2* seedlings. These findings prompt two conclusions: first, that strigolactone signalling is independent of MAX2 in moss and second that the original role for MAX2 was in photomorphogenesis and early development rather than SL perception. Our knowledge of strigolactone perception and signalling pathway is still very fragmentary in bryophytes, and even more so in Charales, but it should become clearer in the future, at least for *Physcomitrella* and *Marchantia*.

5.4.3   SMXL Proteins

The pattern of SMXL gene distribution through the land plants is complex; however, it is clear that SMXL sequences are present in all land plant groups and absent from charophyte algae (Table 5.1). A recent phylogenetic analysis indicates that, unlike many SL biosynthetic enzymes that are typically present in single copies in plant genomes, SMXL proteins show extensive duplication and diversification. The overall trend is one of increasing SMXL diversity with time. Although bryophyte, lycophyte, and monilophyte genomes typically only contain a single variety of *SMXL* gene, seed plants contain between two and four distinct types, with each type exhibiting extensive paralogy; some genomes (e.g. *Arabidopsis*) contain up to eight *SMXL* genes across four major clades. Accordingly, there seems to have been a core *SMXL* gene present in the last common ancestor of all land plants, and this was maintained until the evolution of the seed plants. At this point, gene duplication and neofunctionalisation seem to have increased the variety of SMXL sequences. Most importantly, the class of SMXL proteins specific to strigolactone signalling in angiosperms (SMXL6, SMXL7, and SMXL8 in *Arabidopsis*; D53 in rice) do not arise in land plants until the evolution of the angiosperms (Fig. 5.1) (Walker and Bennett 2017). This innovation neatly coincides with the evolution of bona fide D14 proteins in the seed plants. Notably, and as is the case with many regulatory gene families in plants, these evolutionary duplications and diversifications coincide with the increased developmental complexity observed in seed plants.

To date, no functional information is available for SMXL proteins outside of angiosperms, and therefore the role of SMXL proteins with respect to strigolactone signalling in most land plant groups is unknown. Although MAX2 is unlikely to have a role in strigolactone signalling in moss, it is still possible that moss SMXL proteins are involved in the process, perhaps through direct interaction with KAI2/DDK proteins. In angiosperms at least, D14 is capable of direct interaction with SMXL7 and D53, independently of MAX2. Thus, while MAX2 is absolutely required for strigolactone signalling in seed plants, there is a possibility that a more simplified "noncanonical" signalling mechanism exists in basal land plants that relies upon SMXL proteins but bypasses regulation by MAX2. Since nothing is currently known about how SMXL proteins function in bryophytes, we can only speculate. But suppose that in bryophytes, rather than being degraded, SMXLs are sequestered and rendered non-functional by forming a complex with a KAI2/DDK protein. As outlined above, this relatively primitive mechanism could be improved upon by the introduction of MAX2 to remove SMXL activity through protein degradation rather than simply sequestration.

5.5 Strigolactones in Non-seed Plants: Hormones or Rhizosphere Signals?

A role for strigolactones as a plant hormone in angiosperms is supported by numerous genetic and physiological studies. In many species, strigolactones also act as a signal secreted by the roots into the rhizosphere, which helps to recruit arbuscular mycorrhizal fungi and thereby improve plant nutrient availability (see Chap. 4). Arbuscular mycorrhizal (AM) associations are evolutionarily ancient, forming in all major land plant groups, including the bryophytes. This raises the question of whether the ancestral role for strigolactones was one of hormonal regulation of development, or as a rhizosphere signal. Many non-seed plants lack key strigolactone signalling components, especially D14 and SMXL7-type proteins, which would mean another perception and response mechanism would be needed in such species. Indeed, *M. polymorpha* has also lost the ability to synthesise strigolactones because it has lost CCD8 and MAX1 enzymes (Table 5.1) (Walker and Bennett 2017). If strigolactones were primarily a developmental/hormonal signal in this species, loss of the ability to synthesise this signal would presumably not be tolerated and would be selected against. Perhaps not coincidentally, *M. polymorpha* has also lost the ability to recruit AM fungi, whereas other *Marchantia* species have retained AM associations.

But what about strigolactones in moss? As discussed above, we know that *P. patens* produces carlactone and therefore probably SL-like compounds. We also know that *P. patens* requires carlactone biosynthesis for normal development. However, what if *P. patens* is unusual? The apparent role of strigolactones in this species is to limit plant growth in a quorum sensing-like manner, which could be considered a limited developmental response to what is, in essence, a rhizosphere signal. As *P. patens* has also lost the ability to form AM associations (unlike its close relatives), it is possible that the recruitment of strigolactones as a developmental signal in this species is an independent innovation unique to this species or close relatives. While it is premature to draw firm conclusions, it seems reasonable that the ancestral function of strigolactones was as a means for plants to signal to nearby organisms in the rhizosphere. More research into the functional requirement for strigolactone biosynthesis in a wider range of non-seed plants would help to support or refute this hypothesis.

5.6 Evolution of Strigolactone Receptors in Parasitic Weeds

As outlined in Chap. 3, strigolactones were discovered by virtue of their capacity to stimulate seed germination of root parasitic weed species in the Orobanchaceae, such as *Striga hermonthica*. Several species within the Orobanchaceae are obligate parasites that can cause devastating crop losses. The seed of these species are adapted to long periods of dormancy before detecting root exudates from a nearby host root. In many cases, strigolactones are the key chemical signal that triggers germination. Indeed, measuring seed germination of parasitic weeds is still among the most sensitive methods for detecting the presence of strigolactones in biosamples.

A long sought-after method for controlling parasitic weeds is one of suicidal germination, by which the artificial application of a chemical germination stimulant would trigger depletion of the soil seed bank and render the field safe to sow a crop. Knowledge of the mechanism of strigolactone perception by the parasites would greatly enhance the prospects of suicidal germination approaches, because it would permit the rational design of suitable chemical analogues that are cheaper and easier to synthesise, and more chemically stable, than strigolactones themselves. Our understanding of the strigolactone signalling mechanism in model species has opened up the route to solve this problem.

Perhaps surprisingly, parasitic weeds do not detect host strigolactones using D14 as a receptor but instead use eu-KAI2 homologues. In many species, there has been extensive gene duplication in the eu-KAI2 family, whereas D14 is typically present only as a single gene copy (Conn et al. 2015; Toh et al. 2015). A general pattern has emerged that, like all land plants, these parasites have maintained a conserved eu-KAI2 gene copy that presumably retains its function as a receptor for KL and for overall plant development. But in the plant lineage containing the Orobanchaceae, there are additional KAI2 gene copies that have since diverged and have undergone fast rates of evolution. Several such "KAI2d" paralogues are predicted to have enlarged ligand-binding pockets relative to the eu-KAI2 ancestral copies, suggesting that they have gained the ability to accommodate a larger ligand—such as a strigolactone molecule—and thus have adjusted their substrate preference towards strigolactones. Indeed, biochemical and structural characterisation of several KAI2d proteins has indicated that they can hydrolyse strigolactones with varying affinities (Xu et al. 2018). Furthermore, when these KAI2d proteins are expressed in transgenic *Arabidopsis* plants, they can impart strigolactone-dependent germination upon seed that normally does not respond to strigolactones. Thus, it appears that gene duplication followed by neofunctionalisation of KAI2 paralogues has allowed specific and sensitive control of seed germination in parasitic weed species.

In retrospect, that parasitic weeds have made use of KAI2 receptors to detect strigolactones makes intuitive evolutionary sense. A role for KAI2 in the regulation of seed dormancy and seed germination is clear from the *Arabidopsis kai2* mutant phenotype. It is also clear from genetic studies in *Arabidopsis* that KAI2 operates through the activity of a specific and highly conserved clade of SMXL proteins that are not involved in strigolactone response in most plants but are involved in the KAI2 pathway and the control of seed germination. It is potentially more parsimonious from an evolutionary point of view to adjust the ligand-binding pocket and substrate affinity of a KAI2 receptor, which is already poised to regulate seed germination, than it is to assign D14 as a regulator of seed germination and also to change its specificity for downstream SMXL proteins. Furthermore, there might be strong selective pressure against duplication of *D14* genes, perhaps because additional *D14* copies would interfere with endogenous SL signalling processes within the parasite. Such suppression of *D14* duplication would limit the opportunities for neofunctionalisation. Based on sequence analysis of KAI2d paralogues from various parasitic species, it seems that there are many different ways to change the apparent size of the ligand-binding pocket through amino acid substitutions. Once gene duplication has happened, there is extensive evolutionary space available to explore and—in the case of parasitic weeds—very strong selection pressure for getting it right. It is possible that much of the sequence variation observed among parasitic KAI2d sequences may account for the host preferences of individual parasitic species. In this way, a parasite may be best matched to its host by virtue of having receptors that correspond to the particular strigolactone profile exuded by the host roots. Now that we understand in principle how parasitic weeds detect their hosts, it should be possible to design and identify lead compounds for triggering suicidal germination and thereby help to control this agricultural scourge.

5.7 Conclusions

- The evolution of land plants is associated with increases in developmental complexity, brought about by diversification of gene families and hormone signalling pathways.
- The moss *Physcomitrella patens* and the liverwort *Marchantia polymorpha* are model species for the study of early diverging land plants. Both species are amenable to genetic manipulation and targeted gene knockouts.
- With minor exceptions, the core enzymes for strigolactone biosynthesis via carlactone are present in all land plants, suggesting that strigolactones, or strigolactone-like compounds, are common to all land plants. Strigolactone production in green algae is debated.
- While strigolactones have a clear developmental/hormonal role in angiosperms, the function of strigolactones in early land plants is equivocal. It is possible that strigolactones first served as a rhizosphere signal, perhaps to recruit arbuscular mycorrhizal fungi. Nevertheless, at least in moss, strigolactones regulate protonemal branching and radial expansion.
- Receptor-enzymes of the D14-KAI2-DDK family perceive strigolactones and/or strigolactone-like compounds. These proteins predate land plants and have undergone substantial duplication throughout land plant evolution.
- The F-box protein MAX2 is essential for strigolactone perception in angiosperms, but is not required in moss. The role of MAX2 in strigolactone signalling is an acquired innovation of vascular plants.
- Members of the *SMXL* gene family are also universal in land plants and increase in diversity with organismal complexity. Their function in non-angiosperm species has not yet been established.
- Parasitic weeds in the Orobanchaceae demonstrate evolution in action, with germination responses to strigolactones mediated by duplication and sequence diversification of KAI2 receptors.

Glossary

Bryophytes an informal, paraphyletic group of non-vascular land plants that includes the liverworts, mosses, and hornworts.

Embryophytes all terrestrial plants (including those that are secondarily aquatic), a group that emerged within the streptophytes.

Homologue a gene copy related by descent to another gene copy. Such a relationship is inferred on the basis of sequence similarity. A gene may have homologues within a species or between species or both. Homologues may be orthologues or paralogues.

Lycophytes one of the earliest groups of tracheophytes that have microphyllous leaves (small leaves with a single vein).

Monilophytes the true ferns with megaphyllous leaves (large leaves with multiple, branched veins).

Neofunctionalisation after gene duplication to produce two paralogues, an evolutionary process whereby one paralogue undergoes mutation to create a new function that was not present in the ancestral gene, allowing the second gene copy to retain the original function. In contrast with *subfunctionalisation*.

Orthologue a gene copy that is separated from related sequence by a speciation event. That is, a gene in species A is more closely related to a gene in species B than it is to another gene in species A. The two orthologues arose from a gene duplication event that predated the speciation of A and B. Orthologues often have similar functions in both species.

Paralogue a gene copy that is not separated from a related sequence by a speciation event. That is, two genes in species A are more closely related to one another than they are to similar sequences in species B. Paralogues arise through a gene duplication event that happened recently within one or both species A and B but after they speciated. Paralogues may indicate that there is functional redundancy or, given enough evolutionary time and selection pressure, functional specialisation.

Seed plants also known as *spermatophytes*. Plants that bear seeds, namely, angiosperms (flowering plants) and gymnosperms (conifers and allies).

Streptophytes the collection of all land plants and the immediate sister group to land plants, namely, the charophyte algae.

Subfunctionalisation after gene duplication, an evolutionary process whereby each paralogue adopts a different function from each other, when both functions were previously performed by the single ancestral gene. Thus, each paralogue has now become specialised in function from a more generalist ancestor.

Tracheophytes vascular plants, which conduct water from the roots along vascular strands or tracheids.

References

Arite T, Umehara M, Ishikawa S, Hanada A, Maekawa M, Yamaguchi S, Kyozuka J (2009) d14, a strigolactone-insensitive mutant of rice, shows an accelerated outgrowth of tillers. Plant Cell Physiol 50:1416–1424

Bowman JL, Kohchi T, Yamato KT et al (2017) Insights into land plant evolution garnered from the *Marchantia polymorpha* genome. Cell 171:287–304.e15

Bythell-Douglas R, Rothfels CJ, Stevenson DWD, Graham SW, Wong GK-S, Nelson DC, Bennett T (2017) Evolution of strigolactone receptors by gradual neo-functionalization of KAI2 paralogues. BMC Biol 15:52

Challis RJ, Hepworth J, Mouchel C, Waites R, Leyser O (2013) A role for MORE AXILLARY GROWTH1 (MAX1) in evolutionary diversity in strigolactone signaling upstream of MAX2. Plant Physiol 161:1885–1902

Conn CE, Bythell-Douglas R, Neumann D et al (2015) Convergent evolution of strigolactone perception enabled host detection in parasitic plants. Science 349:540–543

Delaux P-M, Xie X, Timme RE et al (2012) Origin of strigolactones in the green lineage. New Phytol 195:857–871

Gutjahr C, Gobbato E, Choi J et al (2015) Rice perception of symbiotic arbuscular mycorrhizal fungi requires the karrikin receptor complex. Science 350:1521–1524

Jiang L, Liu X, Xiong G et al (2013) DWARF 53 acts as a repressor of strigolactone signalling in rice. Nature 504:401–405

Lopez-Obando M, de Villiers R, Hoffmann B et al (2018) *Physcomitrella patens* MAX2 characterization suggests an ancient role for this F-box protein in photomorphogenesis rather than strigolactone signalling. New Phytol 435:824

Nelson DC, Scaffidi A, Dun EA, Waters MT, Flematti GR, Dixon KW, Beveridge CA, Ghisalberti EL, Smith SM (2011) F-box protein MAX2 has dual roles in karrikin and strigolactone signaling in *Arabidopsis thaliana*. Proc Natl Acad Sci U S A 108:8897–8902

Proust H, Hoffmann B, Xie X, Yoneyama K, Schaefer DG, Yoneyama K, Nogué F, Rameau C (2011) Strigolactones regulate protonema branching and act as a quorum sensing-like signal in the moss *Physcomitrella patens*. Development 138:1531–1539

Scaffidi A, Waters MT, Sun YK, Skelton BW, Dixon KW, Ghisalberti EL, Flematti GR, Smith SM (2014) Strigolactone hormones and their stereoisomers signal through two related receptor proteins to induce different physiological responses in Arabidopsis. Plant Physiol 165:1221–1232

Shen H, Zhu L, Bu Q-Y, Huq E (2012) MAX2 affects multiple hormones to promote photomorphogenesis. Mol Plant 5:224–236

Soundappan I, Bennett T, Morffy N, Liang Y, Stanga JP, Abbas A, Leyser O, Nelson DC (2015) SMAX1-LIKE/D53 family members enable distinct MAX2-dependent responses to Strigolactones and Karrikins in Arabidopsis. Plant Cell 27:3143–3159

Toh S, Holbrook-Smith D, Stogios PJ, Onopriyenko O, Lumba S, Tsuchiya Y, Savchenko A, McCourt P (2015) Structure-function analysis identifies highly sensitive strigolactone receptors in Striga. Science 350:203–207

Walker C, Bennett T (2017) Reassessing the evolution of strigolactone synthesis and signalling, biorxiv.org. https://www.biorxiv.org/content/biorxiv/early/2017/12/03/228320.full.pdf. Accessed 30 Jan 2018

Wang L, Wang B, Jiang L et al (2015) Strigolactone signaling in arabidopsis regulates shoot development by targeting D53-Like SMXL repressor proteins for ubiquitination and degradation. Plant Cell 27:3128–3142

Waters MT, Nelson DC, Scaffidi A, Flematti GR, Sun YK, Dixon KW, Smith SM (2012) Specialisation within the DWARF14 protein family confers distinct responses to karrikins and strigolactones in Arabidopsis. Development 139:1285–1295

Waters MT, Gutjahr C, Bennett T, Nelson DC (2017) Strigolactone signaling and evolution. Annu Rev Plant Biol 68:291–322

Xu Y, Miyakawa T, Nosaki S et al (2018) Structural analysis of HTL and D14 proteins reveals the basis for ligand selectivity in Striga. Nat Commun 9:3947

Yoneyama K, Mori N, Sato T et al (2018a) Conversion of carlactone to carlactonoic acid is a conserved function of MAX1 homologs in strigolactone biosynthesis. New Phytol 111:18084

Yoneyama K, Xie X, Yoneyama K, Kisugi T, Nomura T, Nakatani Y, Akiyama K, McErlean CSP (2018b) Which are the major players, canonical or non-canonical strigolactones? J Exp Bot 111:18084

Zhou F, Lin Q, Zhu L et al (2013) D14-SCFD3-dependent degradation of D53 regulates strigolactone signalling. Nature 504:406–410

Chapter 6
The Chemistry of Strigolactones

Cristina Prandi and Christopher S. P. McErlean

Abstract Focus of this chapter is the chemistry of Strigolactones. The structural features that identify the canonical versus non canonical Strigolactones, as well as the stereochemistry of the Strigol type and Orobanchol type families will be described. A special emphasis will be devoted to the total synthesis of natural Strigolactones as the most reliable and recommended method for successful structure elucidation of these natural products. However, due the complexity of the target molecules and to the high stereochemical control required to retain bioactivity, the synthesis of natural Strigolactones is currently not feasible on a multigram scale for applications in agriculture. In order to study the effect of Strigolactones on various biological processes, model compounds were designed and prepared. Synthetic Strigolactones can be classified into two main categories: (a) analogues, whose structure is very similar to natural SLs; (b) mimics, whose structure is much simpler, but showing a bioactivity resembling that of SLs. A survey of the most promising structures for agricultural applications and the synthetic pathways to access them is herein provided.

Keywords Butenolide · Stereocenter · Enel ether · Canonical and non canonical · Enantiomers · Diasteromers

6.1 Introduction

Strigolactones are a group of small molecules which were first reported after isolation from the root exudates of cotton in 1966 (Cook et al. 1966). These compounds were potent germination stimulants of the parasitic witchweed (*Striga*

C. Prandi (✉)
Department of Chemistry, University of Turin, Turin, Italy
e-mail: cristina.prandi@unito.it

C. S. P. McErlean
School of Chemistry, University of Sydney, Sydney, Australia
e-mail: christopher.mcerlean@sydney.edu.au

© Springer Nature Switzerland AG 2019
H. Koltai, C. Prandi (eds.), *Strigolactones - Biology and Applications*,
https://doi.org/10.1007/978-3-030-12153-2_6

lutea Lour.), which has an economically devastating effect on many important crops. As such strigolactones (hereafter SLs) captured the attention of chemists who sought to (1) determine the structures of these molecules; (2) synthesize the molecules; and (3) synthesize molecules that mimic the biological actions of strigolactones. The following chapter will highlight the progress that has been made in each of these areas, which can collectively be categorized as "the chemistry of strigolactones".

6.2 Strigolactone Structures

Organic molecules possess defined three-dimensional shapes (called stereochemistry), which can be described in terms of atom connectivity, relative stereochemistry, and absolute stereochemistry. In order to aid the readers' understanding of the chemistry of strigolactones, a short discussion about stereochemistry is appropriate, and a particularly useful analogy is to compare organic molecules with your hands.

Molecules are made by connecting atoms to other atoms (atom connectivity) in the same manner that fingers and thumbs are attached to your hand. Relative stereochemistry describes the position of each atom in three-dimensional space relative to another atom, which is like describing the order of your fingers—thumb is next to index finger, is next to middle finger, and so on. But note that this is the same for both your left and right hands and does not distinguish between them. Yet left and right hands are different. For example, you cannot put your right hand into a left-handed baseball glove. This is because your hands are non-superimposable mirror images, which is another way of saying that they possess defined absolute stereochemistry. Chemists say that your left and right hands are "enantiomers". Organic molecules may exist as non-superimposable mirror images (enantiomers), which have the same atom connectivity and relative stereochemistry, but have different "handedness" (Fig. 6.1).

The absolute stereochemistry of molecules is of critical importance in biological systems, because the proteins to which they bind possess only one type of "handedness". It is therefore tremendously important to utilize molecules with known absolute stereochemistry for plant- and animal-based research.

When drawing molecules stereochemistry is shown by lines (╱) which connect carbon atoms in the plane of the page. Wedges (╱) project out of the plane of the page, and dashes (╱) project into the plane of the page.

More than half a century ago, Cook and co-workers isolated the first strigolactones, strigol and strigyl acetate (Fig. 6.2), from root exudates of hydroponically grown cotton plants (Cook et al. 1966). Although ^{1}H NMR, infra-red, and mass spectrometric data were collected for the compounds, the only structural information that could be ascertained was that the molecules contained a butenolide

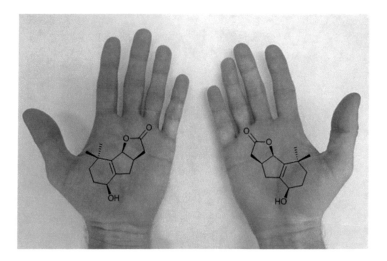

Fig. 6.1 The stereochemistry of molecules

ring, that strigol contained a hydroxyl group, and that the molecules were different than the already known gibberellins. It wasn't until 6 years later that the actual structure of strigol was determined using single crystal X-ray analysis, which unequivocally established both its atom connectivity and relative stereochemistry (Cook et al. 1972; Coggan et al. 1973). Strigol was shown to possess a 6,5,5-tricyclic fragment attached to a butenolide via an enol ether bridge. These are commonly referred to as the A,B,C-tricycle, and the butenolide is termed the D-ring. Importantly, whilst X-ray analysis showed the relative stereochemistry of the molecule, no information about the absolute stereochemistry could be ascertained. Brooks and co-workers were the first to elucidate this crucially important feature of strigol (Brooks et al. 1985). In 1985 they attached an ester with known absolute stereochemistry to strigol and were able to determine that the molecules produced by the cotton plants possessed the ($3aR,5S,8bS,2'R$) stereochemistry. For simplicity, this stereochemical arrangement is commonly referred to as the "strigol-type" stereochemistry. For a long time, it was assumed that all strigolactones possessed the same stereochemistry, but this was later shown to be false. In 2011 (45 years after the isolation of strigol!), Ueno and co-workers completed the chemical synthesis of the naturally occurring strigolactone, orobanchol (Ueno et al. 2011). Although the atom connectivity and relative stereochemistry of the molecule were similar to strigol, the absolute configuration of natural orobanchol was established as ($3aR,4R,8bR,2'R$), which was different to strigol. The different stereochemistry is highlighted in Fig. 6.2 and forms a key division in the family of strigolactones. To date almost 20 strigolactones possessing the common A–D-ring structures have been isolated from various plants. These can be categorized as either "strigol-type" strigolactones or "orobanchol-type" strigolactones (Fig. 6.2).

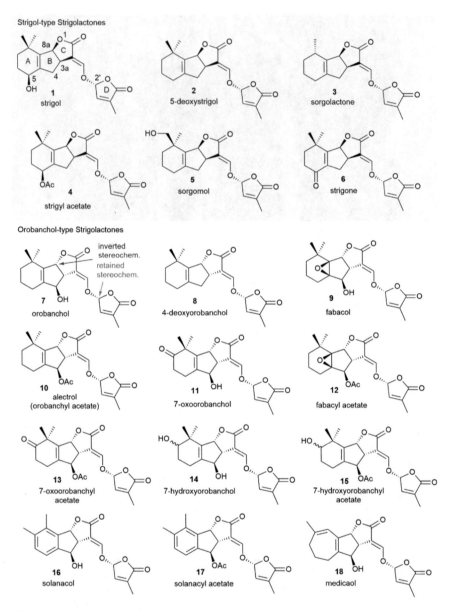

Fig. 6.2 The structure of strigolactones

Strigolactones are produced in minute quantities in plants, meaning that the isolation and structural elucidation of these molecules is exceedingly difficult. The primary technique for identifying new strigolactones is liquid chromatography coupled with tandem mass spectrometry (LC-MS/MS). Plant extracts are typically separated on a chromatographic column, and the eluting molecules are subjected to fragmentation in a mass spectrometer. Because all known strigolactones possess the same butenolide D-ring, the fragment ion with a mass-to-charge ratio of 97, which corresponds to this portion of the molecule, is selectively monitored in sophisticated MS experiments. Molecules that show this ion are collected and analysed.

The pressing need for material to facilitate plant research has provided impetus for chemists to devise methods to synthesize strigolactones. From a purely chemical point of view, the beguiling structures of strigolactones, which feature fused ring systems and an array of oxygen-containing and reactive functional groups, make these molecules exciting targets for organic chemists to display their expertise. Some notable syntheses of strigolactones are detailed below, in the chronological order in which the strigolactone was isolated. Syntheses that control only the relative stereochemistry are termed "racemic" and contain equal amounts of both enantiomers (non-superimposable mirror images of the molecules). Syntheses that result in control of the absolute stereochemistry are termed enantioselective and provide access a single "hand" of these important plant signalling molecules.

6.2.1   Strigol-Type Strigolactones

6.2.1.1   Strigol

As previously mentioned, strigol (**1**) was the first strigolactone identified, being isolated in 1966. The first synthesis of strigol (**1**) was reported in 1974 and allowed access to the sufficient quantities of the molecule for preliminary biological studies. The synthetic strategy employed involved building the A,B,C-ring system and then installing the enol ether bridge and attaching the D-ring. Surprisingly, this general strategy has been followed in every subsequent strigolactone synthesis.

As shown in Scheme 6.1, the commercially available compound citral (**19**) was cyclized to give **20** and **21**, which would become the A-ring of strigol (**1**). Installation of the B-ring of strigol was accomplished by a sequence of reactions that eventually gave the α-keto ester **25**. Synthesis of the C-ring led to a mixture of hydroxylactones **27** and **28**. These were chromatographically separated, and **28**, which possessed the desired relative stereochemistry, was coupled with the bromobutenolide (**29**) to give a mixture of **1** and **30**. Chromatographic separation gave strigol (**1**) with the correct atom connectivity and relative stereochemistry (Heather et al. 1974).

Heather's synthesis of strigol is noteworthy because it represents the first synthesis of any strigolactone molecule. The general synthetic strategy proved to be successful, and material for biological assays was secured, but several limitations were apparent. The synthesis was overly lengthy, which resulted in low overall efficiency; it generated mixtures of products at several stages; and although it controlled relative stereochemistry, it did not control absolute stereochemistry. Subsequent researchers therefore concentrated on developing more efficient strategies to accessing the A,B,C-ring system (Macalpine et al. 1976; Dailey 1987;

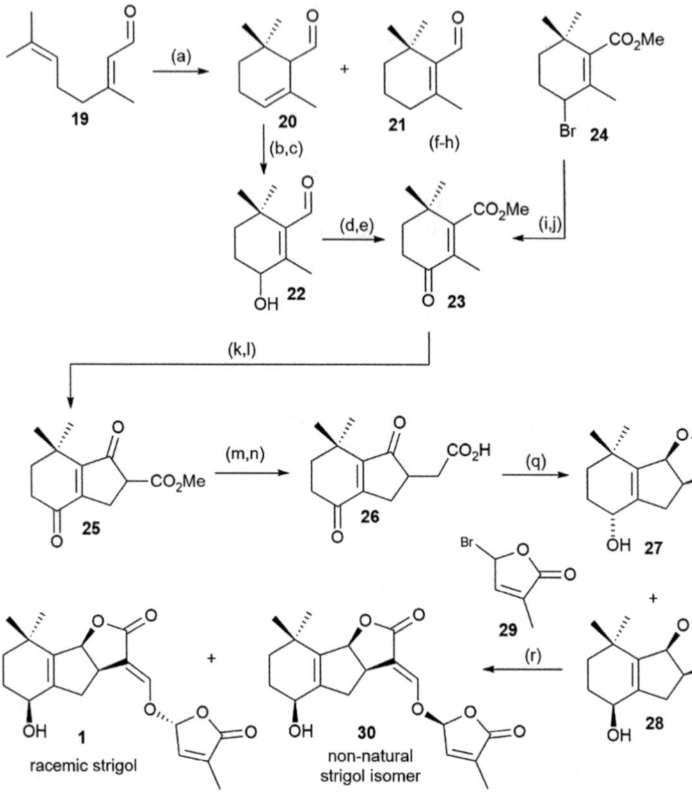

Scheme 6.1 Heather's synthesis of racemic strigol. *Reagents and conditions*: (a) (i) PhNH$_2$, (ii) H$_2$SO$_4$; (b) *m*-CPBA, 90%; (c) pyrrolidine, 73%; (d) CrO$_3$/H$_2$SO$_4$, 50%; (e) MeI, K$_2$CO$_3$, quant.; (f) 5% platinum on carbon, O$_2$, 70%; (g) MeI, K$_2$CO$_3$, 100%; (h) Br$_2$, hν, 100%; (i) Na$_2$CO$_3$, H$_2$O, 81%; (j) CrO$_3$/H$_2$SO$_4$, 98%; (k) NBS, CCl$_4$, 100%; (l) NaCH(CO$_2$Me)$_2$, then reflux, 86%; (m) K$_2$CO$_3$, methyl bromoacetate; (n) AcOH/HCl, 72%; (q) (i) DIBAL, (ii) H$_2$SO$_4$/ H$_2$O, 61%; (r) (i) MeOCHO, NaH, (ii) HCl, (iii) **29**, K$_2$CO$_3$, HMPA

Berlage et al. 1987; Kadas et al. 1996). However, none of those syntheses controlled the absolute stereochemistry of the molecules—they make strigol as an equal mixture of left-hand and right-hand forms (racemic). The preparation of strigolactones as single enantiomers (e.g. only the right-hand form) can conceptually result from three approaches: enantioselective synthesis, resolution of diastereomers, or resolution of enantiomers. Each approach has advantages and limitations, and they will each be discussed below. Due to the importance of accessing strigolactones with known absolute stereochemistry, the remainder of this section will deal exclusively with chemistry that leads to molecules of defined absolute stereochemistry.

The first synthetic access to a single enantiomer of strigol (**1**) was reported by Heather and co-workers in 1976, 2 years after their racemic synthesis, and involved a resolution of diastereomers (Heather et al. 1976).

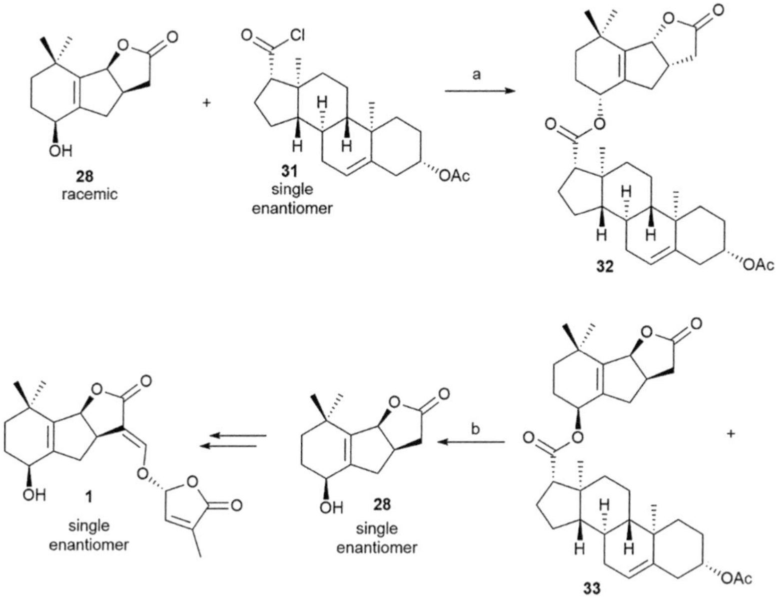

Scheme 6.2 Heather's approach to single enantiomer strigol

As depicted in Scheme 6.2, the hydroxylactone **28** that was synthesized in Scheme 6.1 and existed as a 1:1 mixture of enantiomers was reacted with 3β-acetoxyandrost-5-ene-17β-carbonyl chloride (**31**) to form the corresponding esters **32** and **33**. Since the steroid fragment was a single enantiomer, the two products that were formed were no longer enantiomeric, but instead were diastereomeric. This meant that they could be easily separated by chromatography, and then the steroid ester was cleaved to give **28** as a single enantiomer. Compound **28** was then carried through the same reactions sequence as shown in Scheme 6.1 to give strigol (**1**) as a single enantiomer. At that time that this work was performed, the absolute configuration of naturally occurring strigol was unknown. It would take another 10 years until this was accomplished by Brooks and co-workers (Brooks et al. 1985).

After completing a synthesis that converted α-ionone (**34**) into racemic strigol (**1**) (Scheme 6.3), Brooks and co-workers synthesized a crystalline derivative of strigol using single enantiomer 1-(1-naphthyl)ethyl isocyanate (Brooks et al. 1985). This produced diastereomeric carbamates **35** and **36** which were separated and individually analysed by X-ray crystallography. Having established the stereochemistry of both **35** and **36**, the carbamates were cleaved, and the resulting single enantiomers were compared to naturally occurring strigol. This allowed the absolute configuration of strigol to be determined as (3a*R*,5*S*,8b*S*,2′*R*).

Whilst derivatization of strigol (or an intermediate on route to strigol) with a molecule of known absolute stereochemistry and separation of the products are successful, it is more efficient to separate the enantiomers of strigol directly. This

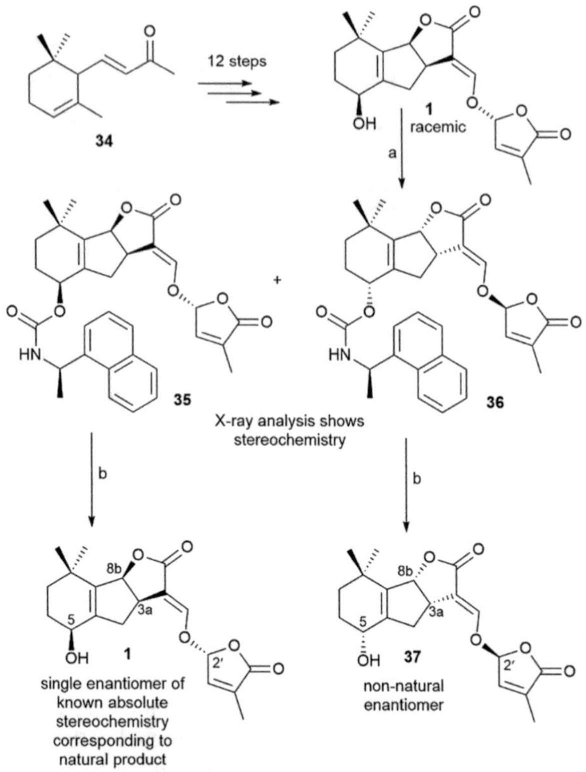

Scheme 6.3 Brook's approach to the single enantiomers of strigol

can be achieved by performing chromatographic separation on a chiral stationary phase (enantioselective chromatography).

Welzel and co-workers used high-performance liquid chromatography (HPLC) on a microcrystalline cellulose triacetate (CTA) stationary phase to separate the enantiomers of compound **28**, enabling him to complete a stereoselective synthesis of strigol (**1**) (Samson et al. 1991). Additionally, Welzel demonstrated that the absolute stereochemistry of the butenolide linkage could be ascertained by examination of the circular dichroism (CD) spectra (Welzel et al. 1999). Compounds containing the naturally occurring 2′R stereochemistry exhibited a negative curve at 270 nm, whereas compounds containing the non-natural 2′S linkage exhibited a positive curve at that wavelength (Scheme 6.4).

Hauck and Schildknecht used HPLC on a CTA stationary phase to separate the enantiomers of strigol (**1**) and determined that the naturally occurring isomer is much more active than the non-natural enantiomer at stimulating striga seed germination (Hauck and Schildknecht 1990). Similarly, Reizelman and co-workers synthesized all of the possible isomers of racemic strigol and separated the individual enantiomers using HPLC on a semi-preparative cellulose carbamate column (Reizelman et al. 2000).

One drawback of approaches that involve separation of enantiomers or diastereomers is that they are restricted to small scales due to the HPLC instruments that are used. To facilitate research into the biological actions of strigol (**1**), larger quantities were needed. This need was met by Mori and co-workers who utilized a chemoenzymatic resolution during their synthesis of strigol (**1**) (Scheme 6.5) (Hirayama and Mori 1999). Treatment of the racemic hydroxylactone **28** with vinyl acetate and lipase AK resulted in a selective reaction with the non-desired enantiomer to give acetate **40** and leave the desired enantiomer **28** untouched. Completion of the synthesis using known procedures allowed Hirayama and co-worker to generate over 1 gram of strigol (**1**) as a single enantiomer (Hirayama and Mori 1999). This remains the only gram-scale synthesis of a strigolactone.

There has only been one reported synthesis of single enantiomer strigol (**1**) that utilizes purely chemical means (which is termed an enantioselective synthesis) (Takahashi et al. 2016). As shown in Scheme 6.6, Kuwahara and co-workers employed a catalytic Corey-Bakshi-Shibata (CBS) reduction of compound **41** in

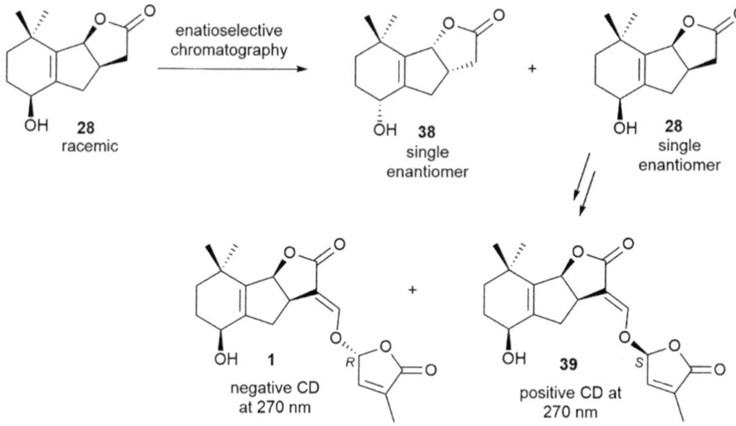

Scheme 6.4 Welzel's approach to the single enantiomers of strigol

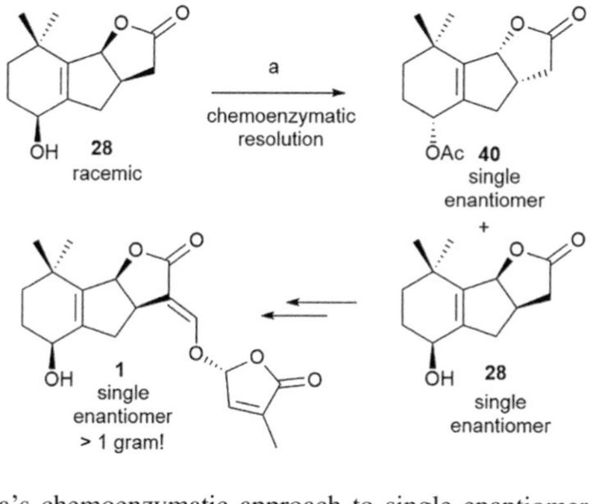

Scheme 6.5 Hirayama's chemoenzymatic approach to single enantiomer strigol. *Reagents and conditions*: (a) vinyl acetate, lipase AK, **40** 52%, **28** 48%

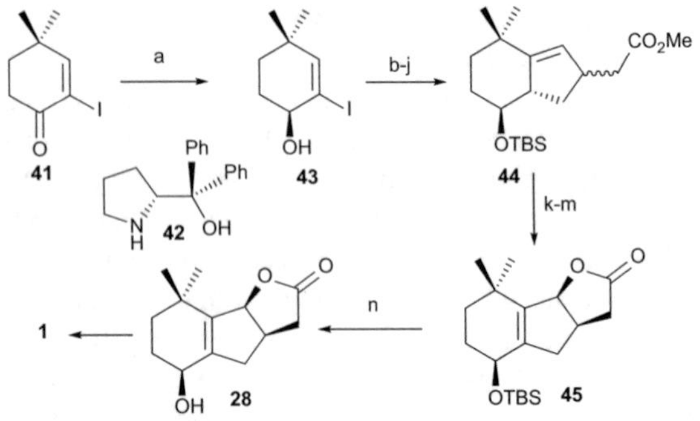

Scheme 6.6 Kuwahara's enantioselective synthesis of strigol. *Reagents and conditions*: (a) **41**, B (OMe)$_3$, PhNMe$_2$•BH$_3$, 92%; (b) H$_2$, Pd/C; (c) *m*-CPBA; (d) TBSCl, 78% three steps; (e) allylMgBr; (f) SO$_3$•Py, DMSO, 67% two steps; (g) vinyl lithium; (h) Grubbs 2nd gen. cat., 97% two steps; (i) Ac$_2$O, DMAP; (j) BF$_3$·OEt$_2$, CH$_2$C(OTBS)OMe, 63% two steps; (k) NaOH, MeOH; (l) NBS; (m) DBU, 23% three steps; (n) TBAF, 76%

which the prolinol **42** dictates which face of the molecule undergoes reaction. This produced compound **43** as a single enantiomer, and the stereochemically defined hydroxyl group was then used to control the stereochemistry of all the other centres during the synthesis.

Despite being reported 50 years after the isolation of strigol (**1**), this enantioselective synthesis is lengthy and suffers from poor overall efficiency, meaning that resolution approaches remain the most practical method of attaining single enantiomer strigol (**1**) for biological applications.

6.2.1.2 Sorgolactone

It is testament to the difficulty of isolating and identifying strigolactones from natural sources that it was a further 26 years after the isolation of strigol (**1**) before another strigolactone was identified. Sorgolactone (**3**) was isolated from *Sorghum bicolor* by Hauck and co-workers in 1992 (Hauck et al. 1992). In contrast to the other strigolactones shown in Fig. 6.2, sorgolactone (**3**) has undergone biosynthetic excision of one of the methyl units on the six-membered ring. Given that several aspects of strigolactone biosynthesis remain unclear, it is unsurprising that the mechanism of this carbon deletion has not yet been elucidated. As the amount of sorgolactone (**3**) isolated was miniscule, it fell to synthetic chemists to confirm the atom connectivity and stereochemistry of the molecule. As frequently occurs in organic chemistry, two research groups independently published syntheses of sorgolactone (**3**) that confirmed its atom connectivity and relative stereochemistry almost concurrently.

Zwanenburg and co-workers reported the first synthesis of racemic sorgolactone (**3**) and used two-dimensional NMR experiments to elucidate the relative stereochemistry (Sugimoto et al. 1997). Mori and co-workers reported a racemic synthesis of sorgolactone (**3**) that involved a crystalline intermediate (Mori et al. 1997). In agreement with Zwanenburg, an X-ray analysis served to unambiguously show that the methyl group pointed to the opposite side of the molecule as the five-membered ring.

The first access to single enantiomer sorgolactone (**3**) was reported by Zwanenburg and co-workers (Scheme 6.7) (Sugimoto et al. 1997, 1998). Racemic compound **46** from their previous synthesis was formylated and coupled with the single enantiomer D-ring surrogate **47** (Thuring et al. 1996). This gave diastereomers **48** and **49** that were separated, and then heating initiated a cycloreversion reaction that unveiled the D-ring. In this manner Zwanenburg was able to establish the absolute stereochemistry of sorgolactone (**3**) as depicted.

Mori and co-workers took a different approach to single enantiomer sorgolactone (**3**) and generated the A,B,C-tricycle as a single enantiomer (Scheme 6.8) (Mori and Matsui 1997; Matsui et al. 1999c). Using (S)-methyl citronellate (**51**) as a starting material ensured that the absolute stereochemistry of the methyl group was fixed. Compound **51** was transformed into the alkyne **52**, which was cyclized and modified to give **53**. Construction of the B- and C-rings in the standard way led to two separable diastereomers, **46** and **54**. Unlike the previous syntheses, compound **46** had defined relative and absolute stereochemistry, so it existed as a single enantiomer. Attachment of the D-ring in the normal way gave single enantiomer sorgolactone (**3**).

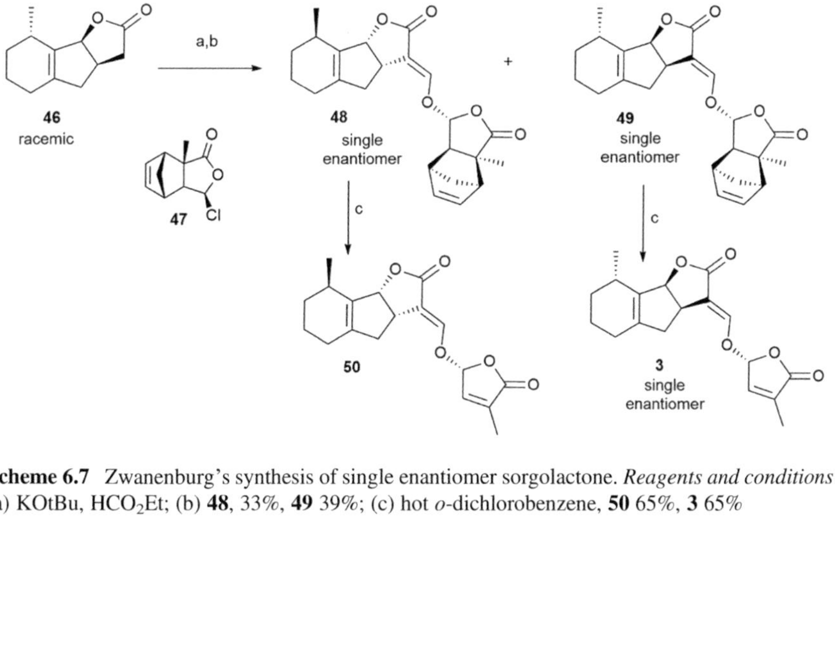

Scheme 6.7 Zwanenburg's synthesis of single enantiomer sorgolactone. *Reagents and conditions*: (a) KOtBu, HCO$_2$Et; (b) **48**, 33%, **49** 39%; (c) hot *o*-dichlorobenzene, **50** 65%, **3** 65%

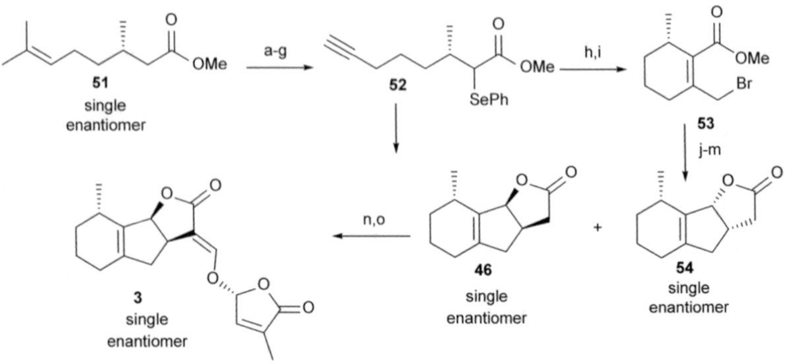

Scheme 6.8 Mori's synthesis of single enantiomer sorgolactone. *Reagents and conditions*: (a) m-CPBA; (b) HIO4; (c) NaBH$_4$, 91% over three steps; (d) TsCl, pyridine; (e) NaI, 82% over two steps; (f) LiCC·EDA, 37%; (g) LDA, PhSeBr, 69%; (h) Bu$_3$SnH, AIBN, benzene, 55%; (i) C$_5$H$_5$N·HBr, pyridine, 52%; (j) NaH, CH$_2$(CO$_2$Me)$_2$; (k) BrCH$_2$CO$_2$Me, 81%; (l) HCl, AcOH, 96%; (m) NaBH$_4$, CeCl$_3$·7H$_2$O, then HCl, **54** 30%, **45** 21%; (n) NaH, HCO$_2$Et; (o) **29**, K$_2$CO$_3$, **3** 42%

6.2.1.3 Deoxystrigol

The chemical history of 5-deoxystrigol (**2**) (called deoxystrigol form here on) runs in opposition to its biological preeminence. In plants, carlactone (**55**) is oxidized to carlactonoic acid (**56**, Scheme 6.15), which undergoes cyclization to give deoxystrigol (**2**) (Scheme 6.9) (Seto et al. 2014). All of the strigol-type strigolactones are thought to be produced by site-selective oxidation processes on deoxystrigol **2** (Matusova et al. 2005). It is somewhat surprising then that deoxystrigol was not identified from a plant source until 2005 (Akiyama et al. 2005), nearly 40 years after its daughter molecule, strigol (**1**).

A number of racemic syntheses have been reported (Reizelman et al. 2000; Akiyama et al. 2005; Frischmuth et al. 1991), but the first enantioselective synthesis of deoxystrigol (**2**) was achieved by De Mesmaeker and co-workers who introduced a new method for construction of the A,B,C-tricycle (Scheme 6.9) (Lachia et al. 2014). Their synthesis began by converting Hagemann's ester (**57**) into the acid **58**. Amide bond formation between **58** and the proline derivative **59** gave compound **60** as a single enantiomer. The proline unit served as a chiral auxiliary that controlled the stereochemistry of the subsequent reactions. Compound **57** was activated to give a ketene-iminium species which underwent immediate intramolecular reaction to give **61**. Having fulfilled its role as a stereo-directing group, the proline derivative was removed to give the four-membered cyclic ketone **62**. Regioselective oxidation gave compound **63** as a single enantiomer, which intersected the previously described racemic synthesis. The D-ring was appended in the standard fashion to give deoxystrigol (**2**) with control over the atom connectivity, the relative stereochemistry, and the absolute stereochemistry.

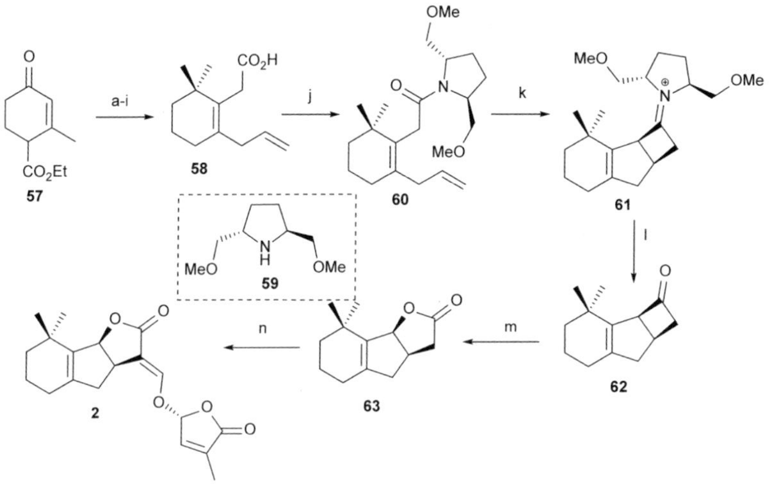

Scheme 6.9 De Mesmaeker's enantioselective synthesis of deoxystrigol. *Reagents and conditions*: (a) NaOH; (b) BrCH$_2$CO$_2$Et, 100% over two steps; (c) NaOH; (d) HCl; (e) toluene, DMF, heat, 100% over three steps; (f) Cs$_2$CO$_3$, BnBr, 86%; (g) MeLi, CuI, Commin's reagent, 100%; (h) allylSnBu$_3$, Pd(PPh$_3$)$_4$, LiCl, heat, 90%; (i) NaOH, 86%; (j) **58**, EDC, HOAt, NEt$_3$, 85%; (k) Tf$_2$O, 2-fluoropyridine, 71%; (m) H$_2$O$_2$, AcOH, 92%; (n) KOtBu, HCO$_2$Et, then **63**, **2** 43%

In 2016 one of the more important papers regarding the chemistry of strigolactones was published by Zwanenburg and co-workers (Zwanenburg et al. 2016b). Whilst previous efforts had ascertained the absolute stereochemistry of strigol (**1**) (Brooks et al. 1985), the assignment of the absolute stereochemistry of deoxystrigol (**2**) was based on Welzel's mnemonic regarding the shape of the CD curve at a specific wavelength (Samson et al. 1991; Welzel et al. 1999). Worryingly, there are examples of molecules containing a D-ring butenolide for which Welzel's mnemonic does not apply (Seto et al. 2014). In order to be confident about the absolute stereochemistry of deoxystrigol (**2**), it was necessary to synthesize all possible isomers, resolve the individual enantiomers, measure the CD spectra, and correlate the stereochemistry to the observed CD curves using an independent technique. This arduous task was successfully undertaken by Zwanenburg et al. (2016b). Using their previously devised synthetic strategy (Reizelman et al. 2000), those researchers synthesized racemic deoxystrigol (**2**) and other stereoisomers. Preparative-scale enantioselective HPLC was then used to isolate the single enantiomers, which are depicted in Fig. 6.3.

Compounds **2** and **64** have the opposite absolute stereochemistry at all positions. This makes them non-superimposable mirror images (enantiomers). In the same manner, compounds **65** and **8** are enantiomers. Compounds **2** and **62** have the same absolute stereochemistry in the A,B,C-rings but have opposing stereochemistry on the D-ring, making them diastereomers. Similarly, compounds **2** and **8** have the same stereochemistry at the D-ring, but opposite stereochemistry on the A,B,C-tricycle, which makes them diastereomers. Zwanenburg obtained an X-ray structure

Fig. 6.3 Zwanenburg's deoxystrigol isomers

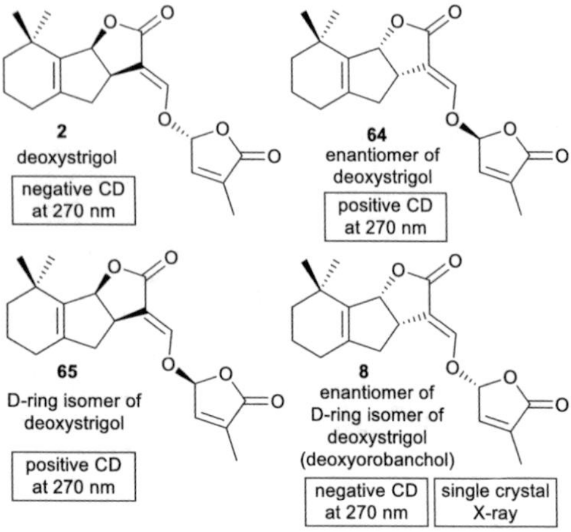

of compound **8**, which unambiguously showed its relative stereochemistry, and measured the CD spectrum, which was negative at 270 nm. If Welzel's mnemonic holds for deoxystrigol, then compound **2** must also have a negative CD curve at 270 nm. Gratifyingly, the measured CD spectrum did show a negative curve at that wavelength. If the D-ring is responsible for the shape of the curves at 270 nm in the CD spectra, then compounds **64** and **65** must display a positive CD curve at that wavelength, even though they have different A,B,C-ring systems. Importantly, **64** and **65** did possess positive curves at 270 nm—Welzel's mnemonic accurately predicted the stereochemistry of the D-ring. The outcome of this is that naturally occurring deoxystrigol either possessed structure **2** or structure **8**, and these diastereomers were easily distinguished by HPLC methods and correlated to De Mesmaeker's synthetic molecule. The outcome of Zwanenburg's efforts is that the absolute stereochemistry of deoxystrigol (**2**) (and by analogy, all strigolactones) was ascertained in an unambiguous manner.

6.2.1.4 Sorgomol

Sorgomol (**5**) was isolated from *Sorghum bicolor* and structurally elucidated in 2008 by Yoneyama and co-workers (Xie et al. 2008a). Racemic sorgomol (**5**) has been synthesized by Takikawa and co-workers (Kitahara et al. 2011). In a landmark piece of research, Sugimoto and co-workers synthesized deuterated deoxystrigol (**2**) and demonstrated that it was converted into deuterated sorgomol (**5**) by the host plant (Scheme 6.10). Furthermore, the oxidation could be inhibited in a dose-dependent manner by the action of uniconazole-P, a known cytochrome P450 inhibitor. This was the first verification of the hypothesized role of deoxystrigol (**2**) as the progenitor of the other strigol-type strigolactones. Although it was not identified in that particular experiment, it is likely that sorgomol (**5**) can be further oxidized and undergoes decarbonylation or decarboxylation to give sorgolactone (**3**).

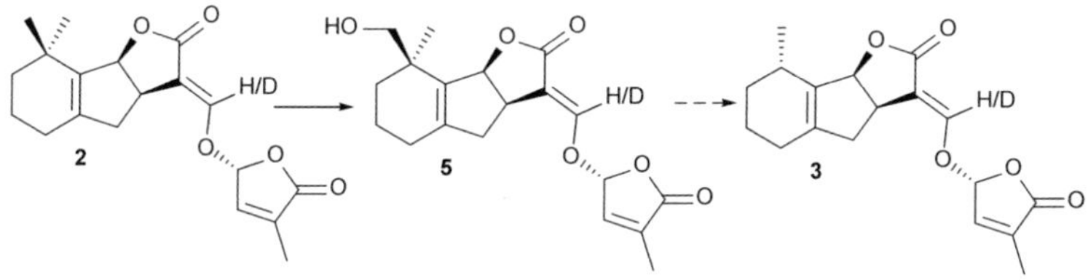

Scheme 6.10   Sugimoto's demonstrated biosynthesis of sorgomol

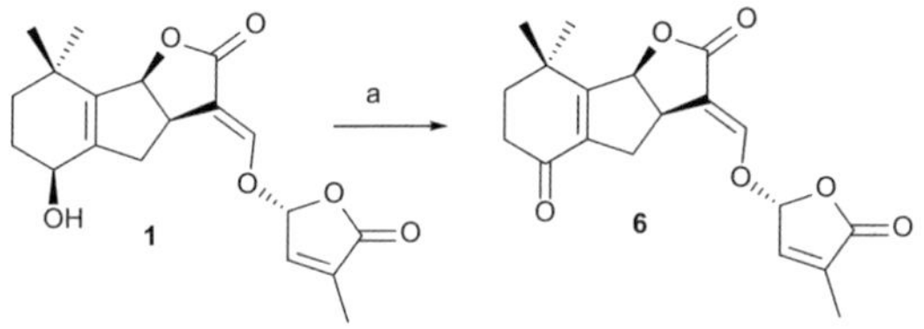

Scheme 6.11   Cook and Welzel's syntheses of strigone. *Reagents and conditions*: (a) MnO$_2$, or pyridinium dichromate, 60%

6.2.1.5   Strigone

Strigone (**6**) was isolated by Yoneyama and co-workers in 2013 (Kisugi et al. 2013). The biosynthetic relationship between strigol (**1**) and strigone (**6**) is clear, with deoxystrigol (**2**) undergoing initial oxidation to strigol (**1**), followed by a secondary oxidation to give strigone (**6**). The first synthesis of strigone (**6**) appeared 41 years before the molecule was isolated from a natural source. As shown in Scheme 6.11, Cook and co-workers oxidized naturally occurring strigol (**1**) (which is a single enantiomer) during the structural elucidation of strigol (**1**) (Cook et al. 1972). Similarly, Welzel and co-workers subjected racemic strigol (**1**) to oxidative conditions and isolated racemic strigone (**6**) (Frischmuth et al. 1993).

6.2.2   Orobanchol-Type Strigolactones

Orobanchol-type strigolactones may arguably prove to be more important for plant-based research than strigol-type strigolactones. This follows from two observations: (1) orobanchol-type strigolactones are more widely distributed across the plant kingdom than their strigol-type counterparts (Ćavar et al. 2015) and (2) the number of orobanchol-type strigolactones isolated is double the number of strigol-type molecules (Fig. 6.2), perhaps reflecting either their increased level of production *in planta* or their increased ability to be transformed *in planta*. Given that the first orobanchol-type strigolactones were correctly identified 45 years after the first strigol-type natural products (Ueno et al. 2011), it is unsurprising that the chemistry of orobanchol-type strigolactones essentially mirrors the work described in the preceding sections.

6.2.2.1 Orobanchol

Orobanchol (**7**) is the most prevalent naturally occurring strigolactone, having been identified in more plant species than any other strigolactone (Ćavar et al. 2015). The molecule was first isolated from a clover species by Yokota and co-workers in 1998 (Yokota et al. 1998). As mentioned in the introduction, it was initially assumed that orobanchol (**7**) possessed the same absolute and relative stereochemistry as strigol (**1**), but this is not the case. Mori and co-workers reported a racemic synthesis of orobanchol (**7**) the year after isolation (Matsui et al. 1999a; Hirayama and Mori 1999). It is important to note that they were in fact trying to generate the non-natural stereochemistry, as the absolute stereochemistry was not known.

It is somewhat perverse that the most influential work on the chemistry of orobanchol-type strigolactones did not involve orobanchol (**7**) at all (Ueno et al. 2011). After completing the racemic synthesis mentioned above, Mori's subsequent efforts to generate orobanchol (**7**) as a single enantiomer resulted in the synthesis of four stereoisomers (**66–69**) (Fig. 6.4), but not the correct structure of orobanchol (**7**) (Hirayama and Mori 1999). It was these molecules that Sugimoto and co-workers compared to naturally occurring orobanchol (**7**) using a combination of chromatographic methods, NMR spectroscopy, and circular dichroism (CD) spectrometry. Compound **67** had the correct atom connectivity and the correct relative stereochemistry, but the opposite absolute stereochemistry of the natural product. Therefore **67** must be the enantiomer of orobanchol (**7**). This ground-breaking work enabled the stereochemistry of orobanchol (**7**) to be unambiguously ascertained and demonstrated for the first time that strigolactones existed as either strigol-type or orobanchol-type molecules.

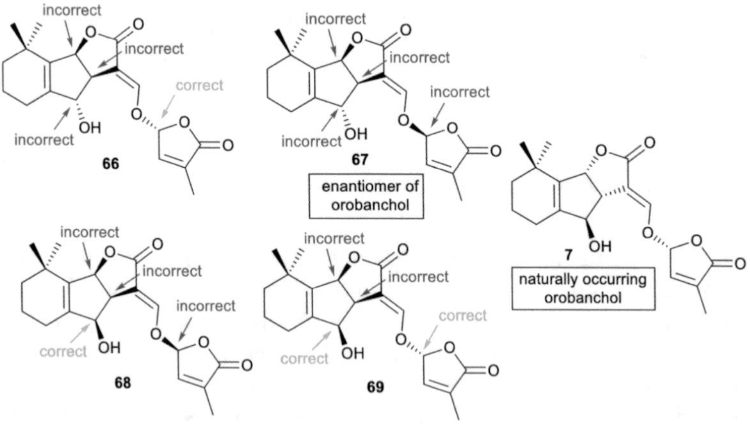

Fig. 6.4 Mori's attempted synthesis of single enantiomer orobanchol

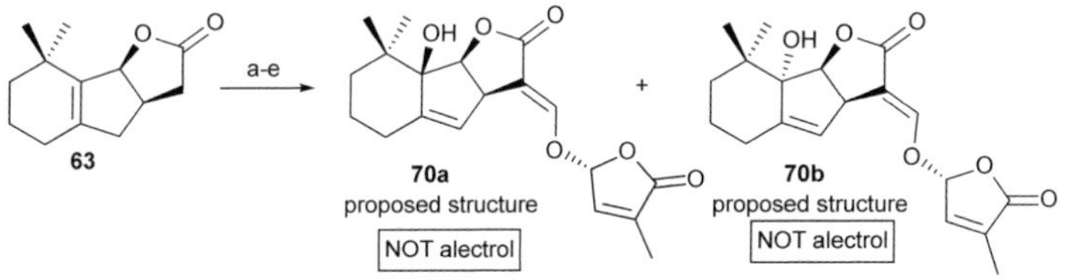

Scheme 6.12 Mori's synthesis of the proposed alectrol structures. *Reagents and conditions*: (a) *m*-CPBA, **70a** 39%, **70b** 60%; (b) Al(OiPr)$_3$; (c) AcOH; (d) NaH, HCO$_2$Et; (e) **29**, K$_2$CO$_3$, **70** 46% over two steps

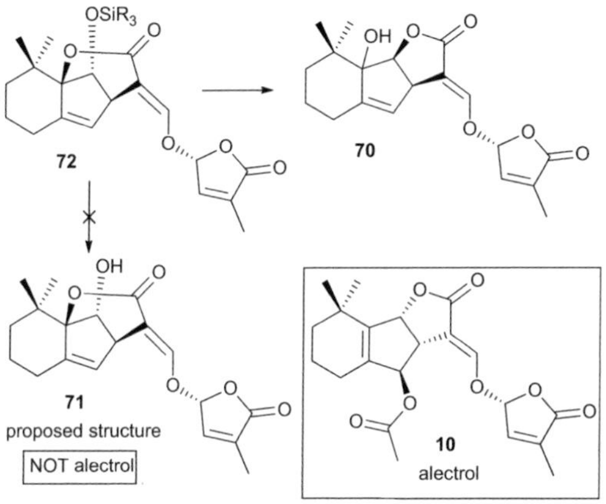

Scheme 6.13 Zwanenburg's attempted synthesis of the proposed alectrol structure

6.2.2.2   Alectrol (Orobanchyl Acetate)

Alectrol (**10**) was first isolated by Müller and co-workers from a cowpea species (*Vigna unguiculata*) (Hauck et al. 1992). After exhaustive chromatographic separations, ~300 µg of this material was isolated and subjected to a range of characterization techniques. Most importantly, the heaviest ion observed in the electron-impact (EI) mass spectrum was m/z = 346 which correlated to a chemical formula of C$_{19}$H$_{22}$O$_6$. On the basis of the spectroscopic data and by comparison to strigol (**1**), Müller and co-workers suggested that alectrol was either structure **70a** or **70b** (Scheme 6.12). Mori and co-workers synthesized **70a** and **70b** starting from the A, B,C-tricycle **63**, but there were significant differences between isolated alectrol and the synthesized material, demonstrating that the hypothesized atom connectivity was incorrect (Matsui et al. 1999b; Mori et al. 1998).

Zwanenburg and co-workers proposed an alternative structure for alectrol **71**, in which the tertiary alcohol from the initially proposed structure was incorporated into the C-ring lactone (Scheme 6.21) (Wigchert et al. 1999). Those authors successfully synthesized the silyl ether **72**, but when they attempted to remove the silyl unit, the molecule spontaneously converted into **70**. This conclusively showed that **71** was not the correct structure for alectrol (Scheme 6.13).

In 2008, Matsuura and co-workers re-isolated alectrol (**10**) from cowpea (*Vigna unguiculata*), and using the mild electrospray ionization (ESI) mass spectrometry, they were able to show that the heaviest ion was m/z = 388 which correlated to a chemical formula of $C_{21}H_{24}O_7$ (Matsuura et al. 2008). In the same year, Yoneyama and co-workers also re-isolated alectrol (**10**) and determined it to be orobanchyl acetate; however, as the stereochemistry for orobanchol had been incorrectly assigned, so too was the stereochemistry of alectrol (**10**) (Xie et al. 2008b). The story reached a conclusion with Sugimoto and co-workers assigning the correct absolute stereochemistry during their structural revision of orobanchol (**10**) (Ueno et al. 2011).

6.2.2.3 Solanacol and Solanacyl Acetate

From a chemical perspective, solanacol (**16**) is one of the more interesting structures among the strigolactones isolated to date, because the A-ring has been oxidized into a benzene ring and one of the methyl groups on the A-ring has migrated to an adjacent position. This suggests that hydroxylated A-ring strigolactones may be intermediates during the biosynthesis of solanacol (**16**), and indeed several such structures have been isolated (see below). As with many strigolactones, the amount of material isolated by Yoneyama and co-workers from a tobacco species was vanishingly small (Xie et al. 2007). This led to errors in the structural assignment. Initially, Yoneyama and co-workers suggested that solanacol possessed structure **73** (Xie et al. 2007). Once again, Takikawa and co-workers undertook the synthesis of this compound only to discover that it did not match the natural product (Takikawa et al. 2009). Those researchers suggested the alternative structure **74**. Finally, the actual structure of solanacol (**16**) was established by total synthesis (Chen et al. 2010) (Fig. 6.5).

As shown in Scheme 6.14, Boyer, Beau, and co-workers transformed the dimethylphenol **75** into the indanone **76** via a number of transformations (Chen et al. 2010). A chemoenzymatic resolution was performed by the action of *Candida antarctica* which gave the alcohol **77** as a single enantiomer. Closure of the C-ring was cleverly achieved using an atom transfer radical addition (ATRA) reaction. As such, **77** was converted into **78** which was treated with copper chloride to give the trichloro-compound **79**. The chlorine atom on the B-ring was replaced with an alcohol, and the chlorine atoms on the C-ring were removed to give the A,B,C-

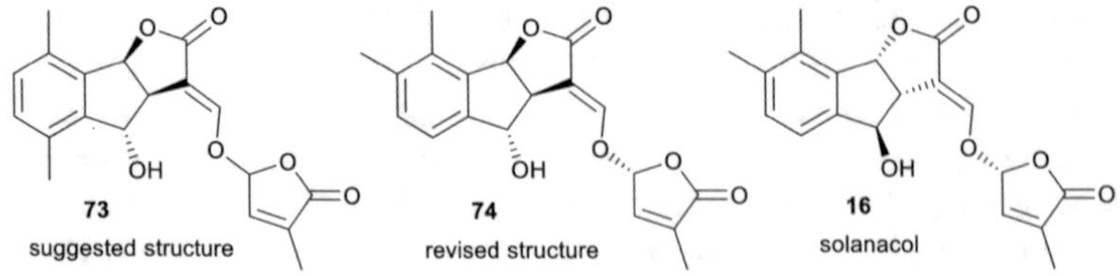

| 73 | 74 | 16 |
| suggested structure | revised structure | solanacol |

Fig. 6.5 Structural revisions of solanacol

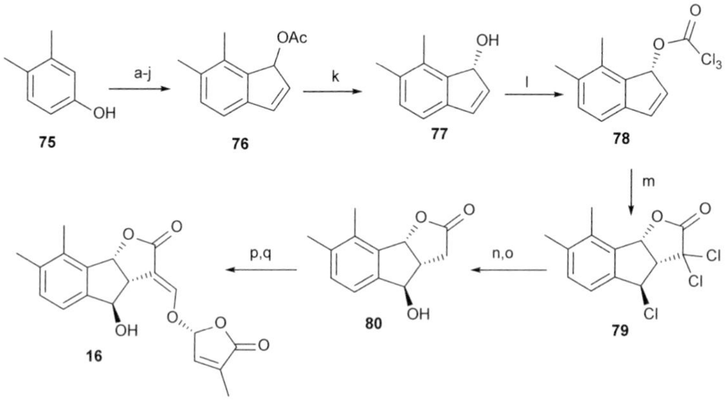

Scheme 6.14 Boyer's synthesis of single enantiomer solanacol. *Reagents and conditions*: (a) Br$_2$, 90%; (b) CH$_2$CHCH$_2$Br, K$_2$CO$_3$, 100%; (c) Et$_2$AlCl, 100%; (d) O3, then Me2S, 79%; (e) H$_2$, Pd/C, Et$_3$N, 100%; (f) Tf$_2$O, pyridine, 74%; (g) CH$_2$CHBF$_3$K, PdCl$_2$, Cs$_2$CO$_3$, PPh$_3$, 98%; (h) CH$_2$CHMgBr, 91%; (i) Grubbs 1st gen cat.; (j) Ac$_2$O, pyridine, 81% over two steps; (k) *C. antarctica*, 50%; (l) (Cl$_3$CCO)$_2$O, pyridine, 98%; (m) dHbipy, CuCl, 83%; (n) H$_2$O, HFIP; (o) Zn dust, NH$_4$Cl, 91% over two steps; (p) *t*-BuOK, HCO$_2$Et; (e) **29**, K$_2$CO$_3$, 38% over two steps

tricycle **80**. Attachment of the D-ring in the standard manner gave solanacol (**16**) with correct atom connectivity, relative stereochemistry, and absolute stereochemistry. Acetylation of **16** gave solanacyl acetate (**17**).

In 2015 Takikawa and co-workers reported an alternative chemoenzymatic approach to single enantiomer solanacol (**16**) (Kumagai et al. 2015).

Whilst orobanchol-type strigolactones are more prevalent than their strigol-type counterparts in the natural environment, they have received decidedly less attention from synthetic chemists. This is due in large to the stereochemical misassignment resulting from a scarcity of isolated material(s). Given that the biological activities of strigolactones are inextricably linked to their three-dimensional shape (stereochemistry) (Scaffidi et al. 2014), the outcomes of experiments involving synthetic orobanchol-type strigolactones prior to 2011 must be viewed with scepticism.

6.2.3 Strigolactone-Related Molecules

Although several aspects of strigolactone biosynthesis remain to be rigorously determined, the broad picture is clear: Carlactone (**55**) is oxidized at C19 to give carlactonoic acid (**56**), which undergoes subsequent oxidation at C18 and stereoselective cyclization leading to either the strigol-type or orobanchol-type strigolactones (Scheme 6.15). The existence of biological machinery capable of achieving these oxidation and cyclization processes suggests that other secondary metabolites resulting from alternative oxidation and cyclization events may exist.

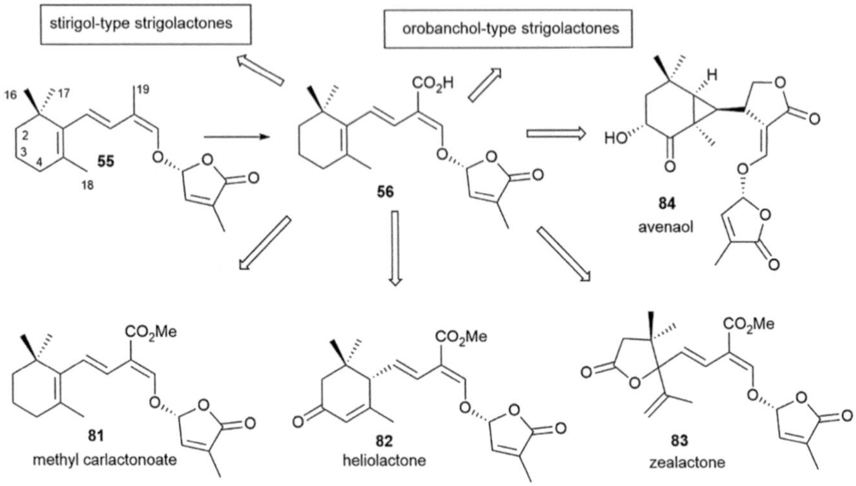

Scheme 6.15 Strigolactone-related molecules

This is indeed the case. Whilst not strictly speaking, strigolactones, a small family of molecules, have been isolated from natural sources that obviously share a common ancestry with the strigolactones, which we have elected to call "strigolactone-related molecules" (and have also been called non-canonical strigolactones) (Scheme 6.15). The biological significance/activity of these molecules is currently under investigation, and this necessitates access to the pure compounds. Therefore, the chemistry of strigolactone-related molecules is a growing area of research.

6.2.3.1 Methyl Carlactonoate

In 2014 Akiyama and co-workers identified a strigolactone-like molecule (SL-LIKE1) in root extracts of *Arabidopsis*, but the metabolite was present in such small amounts that they could not gain structural information about it (Seto et al. 2014). In subsequent work, the structure of SL-LIKE1 was determined to be methyl carlactonoate (**81**) by comparison with synthesized material (Abe et al. 2014). As depicted in Scheme 6.16, the synthesis of methyl carlactonoate (**81**) proved to be very challenging. β-Ionone (**85**) was carried through a nine-step sequence of reactions to produce racemic methyl carlactonoate (**81**). The major drawback of the synthesis was the attachment of the D-ring to compound **86**, which proceeded with only a 3% yield. Clearly, new strategies to improve upon this result needed to be developed.

In 2018, De Mesmaeker and co-workers reported an improved synthesis that also enabled access to single enantiomer methyl carlactonoate (**81**) (Dieckmann et al. 2018). As shown in Scheme 6.17, the major synthetic innovation of this strategy was to unite two fragments to create the central alkene. The racemic D-ring alcohol **87**

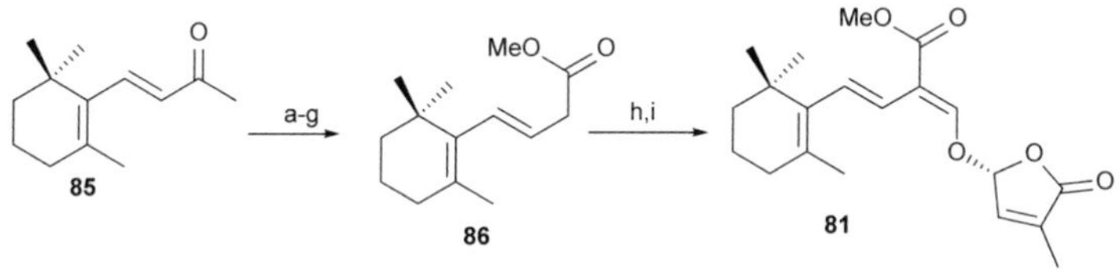

Scheme 6.16 Akiyama's synthesis of racemic methyl carlactonoate. *Reagents and conditions*: (a) NaClO, 96%; (b) Red-Al, 100%; (c) MnO₂, 86%; (d) CH₂ClCO₂Et, NaOMe; (e) NaOH, MeOH; (f) AcOH, 49% over three steps; (g) NaCN,MnO₂, AcOH, MeOH, 29%; (h) NaH, HCO₂Me; (i) **29**, K₂CO₃, 3% over two steps

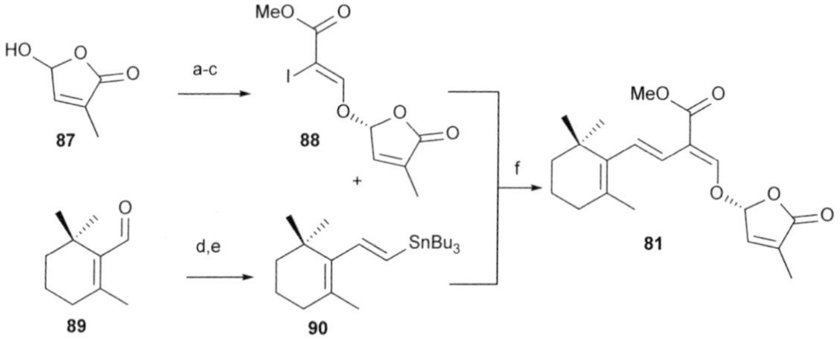

Scheme 6.17 De Mesmaeker's synthesis of single enantiomer methyl carlactonoate. *Reagents and conditions*: (a) HCCCO₂Me, NMM, 77%; (b) *N*-iodosuccinimide, AcOH, then Et₃N, 86%; (c) enantioselective HPLC; (d) TMSCHN₂, LDA, 92%; (e) Bu₃SnH, AIBN, 99%; (f) Pd₂(dba)₃ (7.5 mol%), AsPh₃ (30 mol%), 75%

was added to methyl propiolate and reacted with iodine under basic conditions, and then the products were separated by enantioselective HPLC to give the fragment **88** as a single enantiomer. Separately, β-cyclocitral (**89**) was converted into an alkyne and reacted with tributyltin hydride to give the fragment **90**. Palladium-catalysed cross-coupling united the two fragments and delivered methyl carlactonoate (**81**) as a single enantiomer.

6.2.3.2 Avenaol

The most complex and therefore most synthetically challenging strigolactone-like molecule to be synthesized to date is avenaol (**84**). This molecule was isolated from black oak by Yoneyama and co-workers in 2014 (Kim et al. 2014). The complexity of the molecule is highlighted in Scheme 6.15, with the most challenging feature being the sterically congested three-membered ring.

In a tour de force of synthetic chemistry, Tsukano and co-workers reported the synthesis of single enantiomer avenaol (Yasui et al. 2017). As depicted in Scheme 6.18, the aldehyde **91** was converted through a ten-step sequence into the allene **92**.

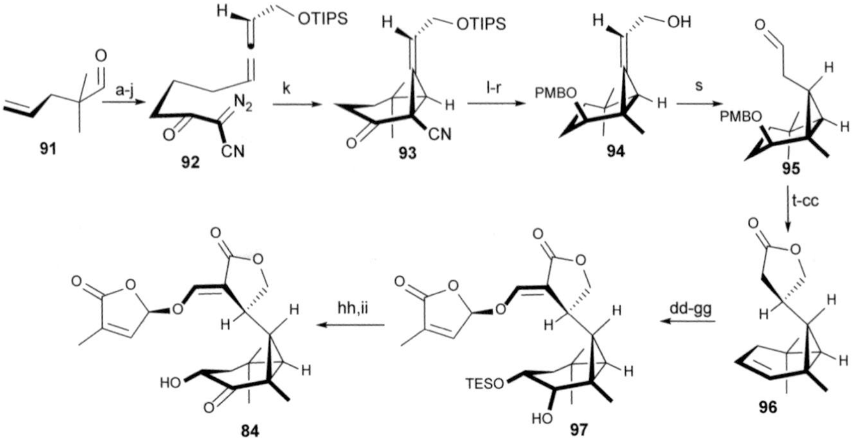

Scheme 6.18 Tsukano's synthesis of single enantiomer avenaol. *Reagents and conditions*: (a) HCCCH$_2$OTHP, BnMe$_3$NOH, 94%; (b) MeI, NaH, 86%; (c) PPTS, MeOH, 92% over two steps; (d) LiAlH$_4$, I$_2$, 83%; (e) TIPSCl, imidazole, 84%; (f) 9-BBN, then NaOH, H$_2$O$_2$, 71%; (g) nor-AZADO, DAIB, 78%; (h) PivCl, 89%; (i) LHMDS, AcCN, 97%; (j) (imid)SO$_2$N$_3$, 87%; (k) Rh$_2$(OAc)$_4$, 84%; (l) NaBH$_4$, CeCl$_3$·7H$_2$O, 95%; (m) PMBCl, NaH, NaI, 97%; (n) DIBAL; (o) NaBH$_4$, 76% over two steps; (p) I$_2$, imidazole, PPh$_3$, 84%; (q) NaBH$_4$, 76%; (r) TBAF, 97%; (s) [Ir(cod)(pyr)PCy$_3$]BAr$_F$, H$_2$, 68%; (t) CH$_2$O, pyrrolidine, EtCO$_2$H, then NaBH$_4$,75%; (u) BH$_3$·THF, NaOH, H$_2$O$_2$, 88%; (v) TsOH, PhSH, 88%; (w) BzCl, Et3N, 86%; (x) TFDO, 96%; (y) MsCl, Et3N; (z) NaCN, 91% over two steps; (aa) DIBAL; (bb) NaOH, then HCl; (cc) TsOH, 49% over three steps; (dd) OsO4, NMO, 63%; (ee) TESCl, imidazole, 76%; (ff) *t*-BuOK, HCO$_2$Me; (gg) **29**, K$_2$CO$_3$, 57% over two steps; (hh) Dess-Martin periodinane, 39%; (ii) HF·Pyr, 97%

A direct cyclopropanation reaction then delivered **93** in which two out of the three stereocentres on the cyclopropane had been installed. After further manipulations, the remaining stereocentre was installed by a highly selective isomerization reaction to give the aldehyde **95**. The success of this isomerization relied on the OPMB group of the precursor **94** blocking one face of the alkene. The aldehyde of compound **95** was used as a handle to build the required five-membered ring, and the OPBM group was removed to give **96**. Attachment of the D-ring in the standard way and reintroduction of oxygen atoms onto the six-membered ring gave **97**. Finally, oxidation to the ketone and removal of a protecting group completed the synthesis of single enantiomer avenaol (**84**) (Scheme 6.18).

Hopefully the preceding discussion on the chemistry of strigol-type, orobanchol-type, and strigolactone-like molecules has served to highlight the chemical strategies that have been employed to access these fascinating natural products. Chemists have deployed myriad reactions to generate molecules with the required atom connectivity, relative stereochemistry, and, increasingly, the desired absolute stereochemistry. But the state of the art is far from mature. In order to ensure that enough material is generated for large-scale applications (possibly even field trials), new strategies for the synthesis of strigolactones must be developed. In the meantime, more synthetically accessible molecules that mimic the biological activities of strigolactones continue to be used, and these form the focus of the next section.

6.3   Analogues and Mimics

Important challenges related with the use of plants as a source for identification of bioactive compounds are related with the accessibility of the starting material. As highlighted in the previous paragraph, often the available amount of natural products is low. Although many natural strigolactones have already been isolated and characterized, available compound quantities are often insufficient for testing for a wide range of biological activities. Whilst small amounts of plant material are usually required for an initial bioactivity evaluation, much larger quantities of pure compounds are needed for a comprehensive characterization of the activity of each constituent. Furthermore, limited availability becomes even more problematic when a bioactive plant-derived natural product is identified to have a very promising bioactivity and becomes a lead for infield applications. The problem of sustainable supply still frequently occurs when dealing with natural compounds. Besides the accessibility of the plant material, also its quality is of great importance. Available plant material often varies on quality and composition, and this can hamper the assessment of its effects. In the case of strigolactones, the chemical composition is not only dependent on species identity but also on soil composition, nutritional aspects, climate, extraction methods, processing, and storage conditions. Moreover, during extraction, as well as during the isolation processes, transformation and degradation of compounds can occur. Further complications related to the resupply of bioactive natural products arise from the fact that strigolactones are more likely to have complex chemical structures with numerous chiral centres, which hampers the development of methods for total synthesis in large scale. In contrast, bioactive leads originating from synthetic libraries are usually comparably easy to generate and modify using simple chemical approaches. Further difficulty is set by the fact that determination of the precise molecular mechanism of action of SLs is a challenging task. There are now 25 natural SLs identified, whose synthesis is tricky, time-consuming, and expensive and is currently not feasible for applications in agriculture. However, it should be stressed that the total synthesis of SLs is the most reliable and recommended method for successful structure elucidation of these natural products. Naturally occurring SLs have a too complex structure for synthesis on a multigram scale. In order to study the effect of SLs on various biological processes, model compounds were designed and prepared. A prerequisite is that SL analogues should have a (much) simpler structure than natural SLs, but at the same time, they should retain their bioactivity. Synthetic SLs can be classified into two main categories, (a) **analogues**, the structure of which is very similar to canonical natural SLs, and (b) **mimics**, the structure of which is much simpler, but they maintain a bioactivity similar to that of SLs (Fig. 6.6).

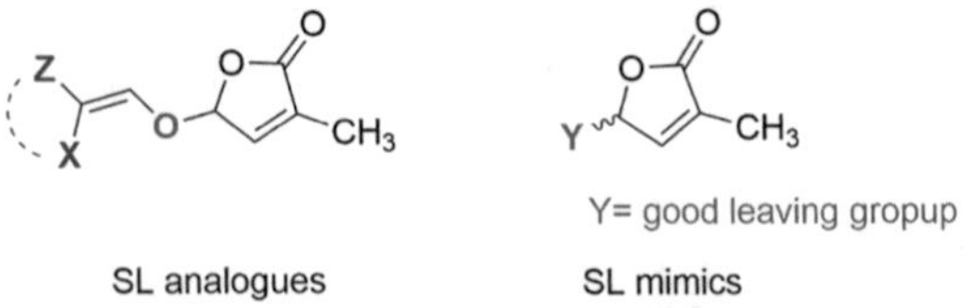

Fig. 6.6 Definition of SL analogues and mimics

SL analogues   SL mimics

Y= good leaving gropup

6.3.1 Analogues

The design and synthesis of SL analogues stems from extensive SAR (structure-activity relationship) which exhaustively assed the bioactiphore of the molecule, in other words which structural feature of the compound is essential to retain bioactivity. As reported in Fig. 6.7, the SL skeleton presents a number of functional groups.

Both C- and D-rings are lactones; the D-ring is an unsaturated lactone known as **butenolide** (see inset). The bridge connecting C- and D-rings is an enol ether; C2' at the junction between the bridge and the D-ring is an acetal.

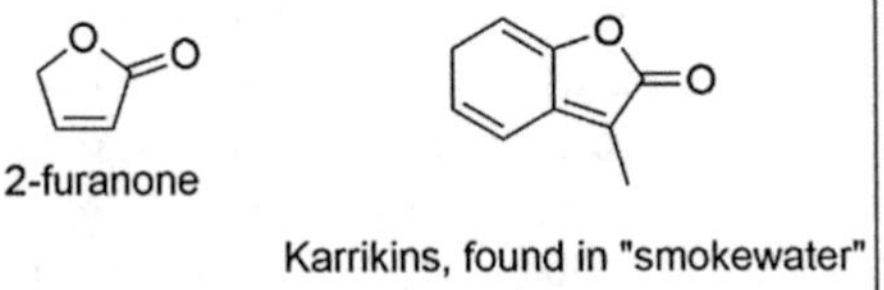

2-furanone

Karrikins, found in "smokewater"

*Butenolides are a class of lactones with a four-carbon heterocyclic ring structure. They are sometimes considered oxidized derivatives of furan. The simplest butenolide is 2-furanone, which is a common component of larger natural products and is sometimes referred to as simply "butenolide". A common biochemically important butenolide is ascorbic acid (vitamin C). Butenolide derivatives known as **karrikins** are produced by some plants on exposure to high temperatures due to brush fires. In particular, 3-methyl-2H-furo[2,3-c]pyran-2-one was found to trigger seed germination in plants whose reproduction is fire-dependent.*

Two α,β-unsaturated ketones are presented and highlighted in blue; both of them can in principle behave as **Michael acceptors** (see inset) and react with nucleophiles. Stereochemistry is also a challenging issue in the synthesis of SLs. As emphasized in Fig. 6.7, right 3 stereocentres and a double bond can give up to 16 different stereoisomers; the number can grow if additional stereocentres are present on rings A and B (see natural SLs in Fig. 6.2 and discussion therein) (Fig. 6.8).

Due to the scarce availability of natural SLs, since the very beginning of the research on this class of hormones, a synthetic SL known as GR24 has been proposed, and very quickly it became the universal standard used as a reference compound for measuring bioactivity. GR24 was initially conceived as an aromatic derivative of strigol (Fig. 6.2); it can be obtained on a multigram scale as a mixture of stereoisomers; recent issues about the observation that different isomers of GR24

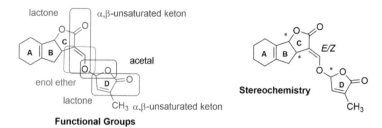

Fig. 6.7 Structural complexity of SLs, functional groups on the left and stereochemistry on the right

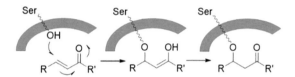

Fig. 6.8 Proposed mechanism of perception for Michael acceptors

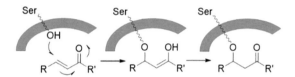

Fig. 6.9 Two families of SLs, strigol-like and orobanchol-like. 5DS as apex stays for same configuration of 5-deoxystrigol, 4DO for 4-deoxyorobanchol

may induce effects that are not related to SLs (Flematti et al. 2016) made the use of pure enantiomers compulsory for a clear-cut interpretation of the biological effects. To simplify the identification of the various stereoisomers, the stereochemistry is related to natural SLs as a consequence GR24[5DS] is the GR24 stereoisomer showing the same configuration as 5-deoxystrigol, GR245[4DO] the same configuration as 4-deoxyorobanchol. These names do not respect the standard IUPAC rules but are now widespread and generally accepted in most scientific journals. Stereoisomers represented in Fig. 6.9 all show R configuration at C2', remarkably in all natural SLs, is invariably the same, namely, R. Enantiomers with S configuration at C2' are much less active. Enantioselective syntheses are now available (Bromhead and McErlean

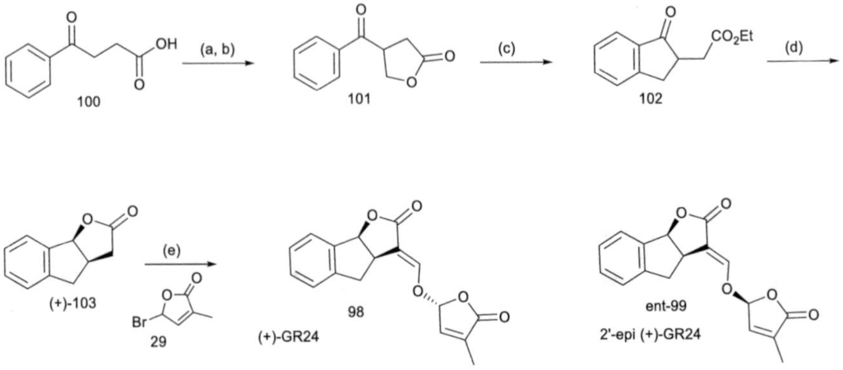

Scheme 6.19 McErlean synthesis of (+)-GR24. Reagents and conditions: (a) HCHO, NaOH (aq); (b)HCl; (c) H_2SO_4, 100 °C, AcCl, EtOH; (d)(S,S)-RuTsDPEN (4 mol%) HCO_2H, $^iPr_2NEt$, then PPTS, 90%, 92% ee to 99% ee after crystallization; (e) HCO_2Me, KOtBu, then **29**

2017) and allow to obtain pure stereoisomers whose bioactivity can be then unequivocally assessed.

Enantioselective synthesis of (+) GR24 **98** has been reported (Scheme 6.19) (Bromhead et al. 2015). The key steps are an intramolecular Ritter reaction to access indanone **102** and a lactonization catalysed by Noyori's (S,S)-RuTsDPEN catalyst to enantiopure (+)-**103**. Remarkably the use of (R,R)-RuTsDPEN gave (−) **103** and paves the way to obtain the other two stereoisomers of GR24.

The gold standard GR24 is routinely used as a positive control in plant-based assays; however it has been recently demonstrated that for seed germination, (−)-4-OH-epi-GR24 **109** is more potent than the parent compound. A biomimetic approach to access the single active enantiomer has been proposed by McErlean according to the synthetic sequence reported in Scheme 6.20 (Morris and McErlean 2016).

A selection of available analogues will be herein presented and classified according to the following structural modifications:

(a) Modification at the ABC core
(b) Modification at the D-ring
(c) Modification at the enol ether

(a) Modification at the ABC core

Within the family of GR derivatives (GR24 being the best-known representative), the concept of designing simpler structures retaining bioactivity led to the synthesis of GR5 and GR7 (Zwanenburg et al. 2016c). This latter was used as a suicidal germinating agent against *S. asiatica*. Worth mentioning is the Nijmegen-1 **115** (Fig. 6.10), which can readily be obtained from simple starting materials in a few synthetic steps, and the germinating activity of which is comparable with that of GR24, as shown in Fig. 6.10 (Zwanenburg et al. 2016c).

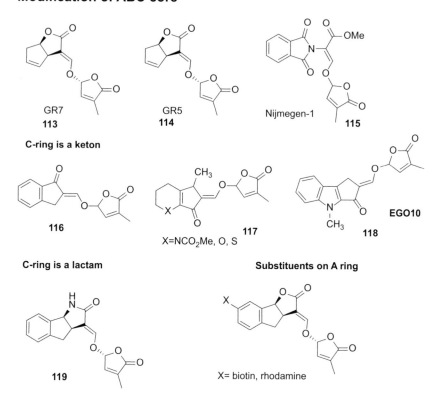

Scheme 6.20 Enantioselective synthesis of 4-hydroxy-GR24 isomers. Reagents and conditions: (a) 159 (20 mol%), 151, Cs$_2$CO$_3$, THF, RT, 38%. (b) (S,S)-RuTsDPEN (4 mol%), HCO$_2$H, DIPEA, DMF, RT, (−)-7 49%. (c) PPh$_3$, DIAD, BzOH, THF, RT, then, K$_2$CO$_3$, MeOH, (−)-8 83%, (+)-8 83%. (d) HCO$_2$Me, t-BuOK, then K$_2$CO$_3$, 29, DMF, RT, (−)-4 21%, (−)-epi-4-19%, (+)-4, 26%, (+)-epi-4 27%. (e) CeCl3·7H2O, NaBH$_4$, EtOH, RT, 68%

Modification of ABC core

GR7
113

GR5
114

Nijmegen-1
115

C-ring is a keton

116

X=NCO$_2$Me, O, S
117

EGO10
118

C-ring is a lactam   **Substituents on A ring**

119

X= biotin, rhodamine

Fig. 6.10 Selected examples of SL analogues with modification at the ABC core

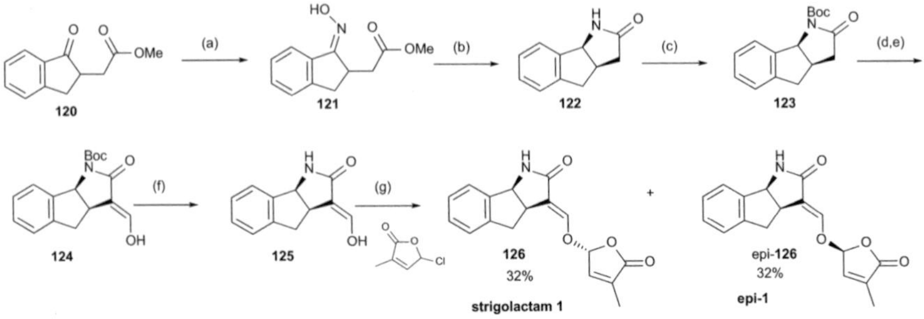

Scheme 6.21 Synthesis of strigolactam-1. Reagents and conditions: (a) NH$_2$ OH·HCl, NaOAc, MeOH, reflux, 88%; (b) Zn, AcOH, 70 °C; then aq NaHCO$_3$; (c) Boc$_2$O, DMAP, Et$_3$N, CH$_2$Cl$_2$, quant.; (d) tert-butoxybis(dimethylamino)methane, neat, 110 °C, 12 h; (e) 1 M HCl, THF, 2 h, 80%; (f) TFA, CH$_2$Cl$_2$, 0 °C, 86%; (g) t-BuOK, THF, then 15, rt., 2 h

SL analogues in which the C-ring has been replaced by a lactone as **116–118** are also available (Prandi et al. 2014) as well as analogues in which the C-ring is a lactam **119** (Fig. 6.10 and Scheme 6.21) (Lachia et al. 2015).

Lactam **122** was obtained in multigram scale from oxime **121**. The lactam **122** was then protected with the Boc group in good yield. A two-step procedure via the formation of the enamine with tert-butoxybis(dimethylamino)methane (Bredereck reagent) and subsequent hydrolysis led to the enol **124** in very high yield. The Boc protecting group was cleaved by treatment with TFA in dichloro- methane at 0 °C for 15 min (longer reaction time led to partial decomposition of the product and low yield). Finally, the enol was deprotonated with potassium tert-butoxide and alkylated with chlorobutenolide. The strigolactam 1 was obtained as a mixture of diastereo-isomers, which were separated by column chromatography (**126** and epi-**126**).

EGO10 **118** is an indolyl-derived SL readily prepared in three steps from available reagents, and it is used as plant hormone in the regulation of shoot branching (Bhattacharya et al. 2009). In addition, SAR studies demonstrated that substituents on the A-ring do not affect the bioactivity of the compounds. Labelled SL analogues with fluorescent probes on the A-ring (rhodamine, fluorescein, BODIPY) have been synthesized and used for investigations on the perception mechanism (Lace and Prandi 2016).

(b) Modification at the D-ring

Modifications at the D-ring mainly consist in the presence of additional sub-stituents in 2′ **128** (Mwakaboko and Zwanenburg 2016), 3′(Boyer et al. 2014) **127**, **129** or in the change of the functional group from lactone to lactam **129** (Fig. 6.11) (Lombardi et al. 2017).

The introduction of extra methyl in 2′ and/or in 3′ does not hamper the biological activity, which in the case of hormonal is even enhanced (Boyer et al. 2012). Strigo-D-lactams have been synthesized by building the D-lactam with a RCM using HG catalysts of II generation (Scheme 6.22).

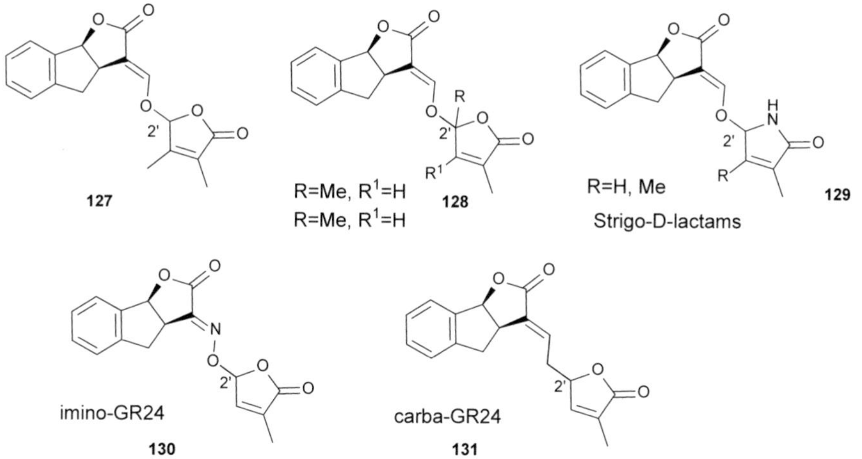

Fig. 6.11 Selected examples of SL analogues with modifications at the enol ether bridge and at the D-ring

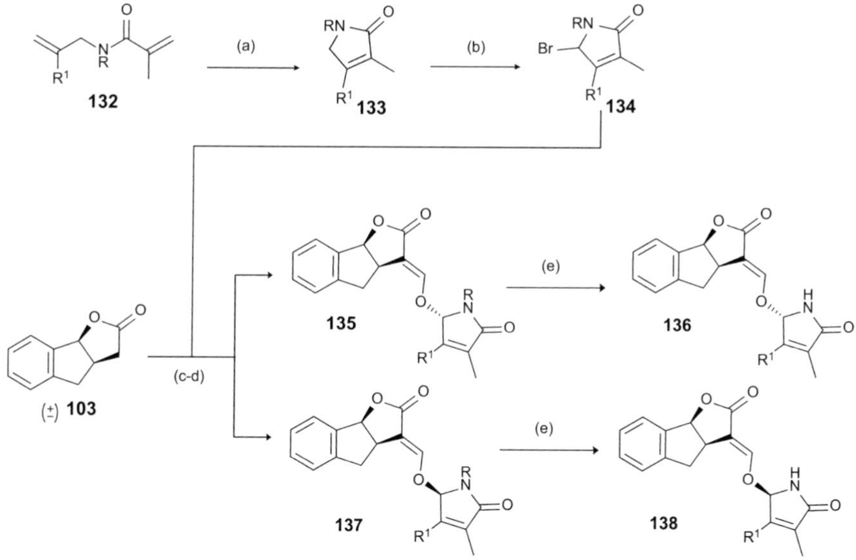

Scheme 6.22 Synthesis of strigol-D-lactam-1. Reagents and conditions: (a) Hveyeda-Grubbs II, PhMe 80 °C, 14–32 hs; (b) NBS, CCl₄; (c) *t*-BuOK, HCOOEt, DME from 0 °C to rt.; (d) lactam X, DME from 0°c to rt.; (e) TFA, DCM, 0 °C, 10 min

Strigo-D-lactams have been tested both as germination inducers and as plant hormones with an *in planta* bioassay, but they proved to be completely inactive at physiological concentrations. This data brought evidence to the hypothesis that the D-lactone ring in SLs plays a crucial role in the perception mechanism.

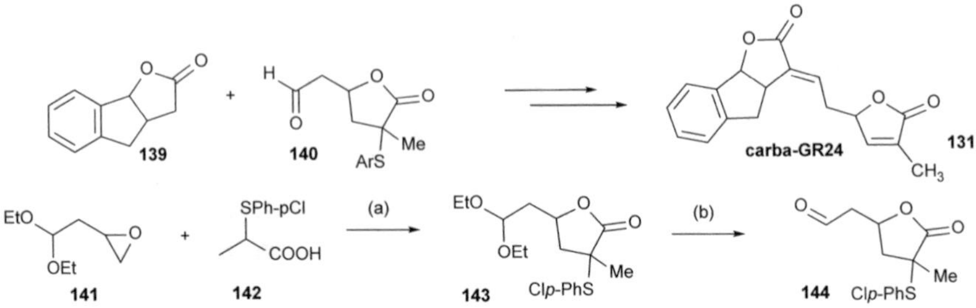

Scheme 6.23 Synthesis of carba-GR24. Reagents and conditions: (a) LDA (2 equiv), aqueous tartaric acid, SiO$_2$, Δ; (b) 0.5 N HCl, Δ

(c) Modification at the enol ether

In designing new analogues with germination capabilities, the replacement of an oxygen by another heteroatom led to two successful examples of such an isosteric replacement, namely, imino SL analogues (Kondo et al. 2007) **182** and carba-GR24 **183** (Fig. 6.11) (Thuring et al. 1997). Whilst carba-GR24 is completely inactive, imino analogues retained some activity. These findings demonstrate that it is not always essential to have the Michael acceptor of the C–D-ring junction moiety which has been proposed to react with nucleophilic species presented at the target site to enhance the activity.

Carba-GR24 has been synthesized according to the synthetic sequence herein reported.

The key step in the synthesis of carba-GR24 proceeds by coupling of tricyclic lactone **139** with an appropriate latent butenolide fragment (Scheme 6.23). It is essential to use this protected D-ring synthon, because the corresponding butenolide is too unstable to survive the coupling conditions. The D-ring synthon **140** was prepared as reported in Scheme 6.23 and used in an aldol reaction with tricyclic **139** followed by elimination of the phenylthio group. The biological data indicate that isosteric replacement of oxygen by carbon causes complete loss of biological activity.

6.3.2 Mimics

The so-called SL mimics are compounds lacking the ABC scaffold but retaining the D-ring connected to an additional group by means of an ether or ester functionality. The term "mimic" comes from the observation that these compounds mimic SL activity. Owing to their simpler structures, retaining high activity, they can be considered to be promising candidates for agricultural applications. One group of mimics with seed germination stimulatory activity show an aryloxy substituent at C-5 as **145** and were named "debranones" (furanones showing debranching activity) because the main activity profile is the inhibition of shoot branching (Fukui et al.

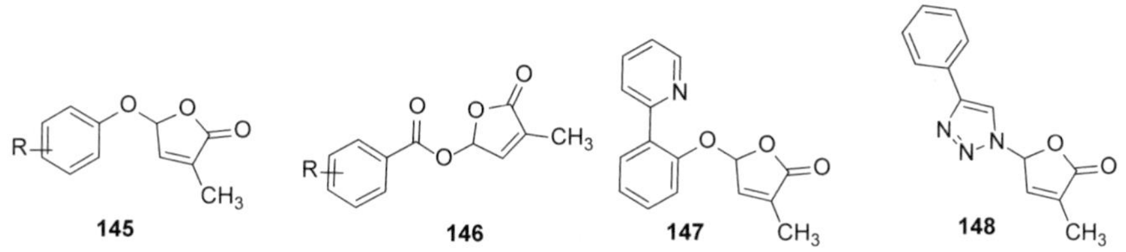

Fig. 6.12 Selected examples of SL mimics

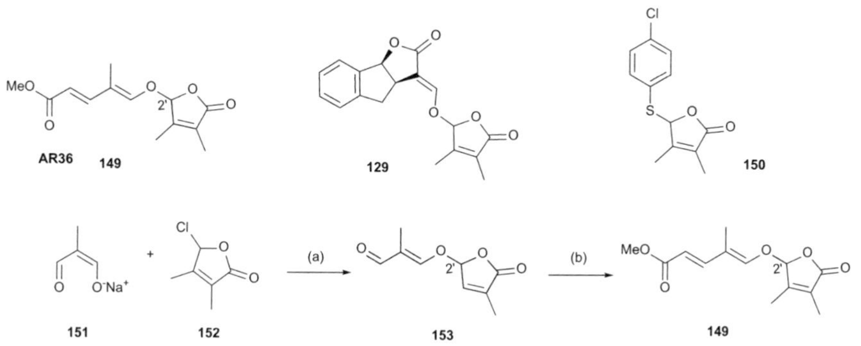

Scheme 6.24 Synthesis of AR36

2017). The second group of SL mimics have an aryloxy substituent at C-5 of the D-rings **146** (Fig. 6.12) (Zwanenburg et al. 2013). These SL mimics are moderately active as germination stimulants towards *S. hermonthica* seeds, but remarkably active in the case of *Orobanche cernua* and *Phelipanche ramosa* seeds. Other recently reported mimics are heterocycle derivatives (Oancea et al. 2017; Dvorakova et al. 2017).

One of the aims in the design and synthesis of new analogues and mimics of SLs is to be able to differentiate the impacts of these compounds on the various target systems, exogenous as parasitic plants and AM fungi and endogenous in their role as plant hormones. A remarkable study in this direction has been conducted by Boyer et al. (2014) (Scheme 6.24).

Synthetic SLs with different structural features have been compared for their activity as parasitic seeds germination inducers, as hyphal branching stimulants in AM fungi and as inhibitors of shoot branching in pea. The study clearly indicated that ABC core is necessary for the hyphal proliferation activity in AM fungi, and the extra methyl at C3′ is crucial for specific endogenous activities. This is contrary to the effect on the germination activity (Zwanenburg et al. 2013; Zwanenburg and Pospisil 2013) where the response depends both on the part of the molecule attached to the D-ring (see Nijmegen and butenolides series) and/or on the parasitic species. Besides, the ABC part of SLs can be replaced by a phenylthio group (**150**) or an unsaturated acyclic carbon chain (AR36, **149**) without changing the bioactivity for the shoot branching inhibition. This is a nice example of how a suitable molecular design can help in separating beneficial from detrimental effect. Up to date there are consistent SAR data indicating that exogenous and endogenous can be clearly dissected; unfortunately the structural requirements to distinguish effects on fungi from those on PP are still to be elucidated.

6.3.3 Inhibitors

Owing to the role of SLs as multifunctional molecules, the search for simple agonists or antagonists may also play a role in both basic research and agricultural applications. Given that most of the enzymes involved in the biosynthesis of SLs are known, biosynthesis inhibitors have been identified and successfully applied. However, the search for perception inhibitors is still in its infancy. To date, all the SL agonists identified show a D-ring or derivative, the only exceptions being the cotylimide (CTL) compounds (Nakamura and Asami 2014), the structure of which does not involve a D-ring. The identification of suitable inhibitors may allow a fine control and tuning of SL effects.

6.4 Applications

After the identification of SLs as phytohormones, intense scientific activity has provided insights into the multiple plant traits that are controlled by the hormonal action of SLs. In general, SLs contribute to plant adaptation in poor soils, and many scientific papers have proposed the use of SLs in agricultural soils with the aim of increasing crop productivity. However, the impact of SLs on the indigenous soil microbial community is unknown. A further chapter in the SL "story" would be to ensure that the use of SLs to enhance crop performance is safe for soil life. Biodegradability of lead compounds, through studies of molecular stability in aqueous media at different pH, their photostability, and the identification of byproducts would be highly desirable. The proven lability of SLs ensures minimal SL persistence in soil and prevents SL accumulation. However, whether SL hydrolysis products influence soil microorganisms (structure, abundance, and function of soil microbial communities) still needs to be investigated. Owing to the availability of only small amounts of SLs, studies of off-target effects have received only limited attention. Among these, some tests considered the use of synthetic GR24 at concentrations up to $8.5 \ 10^{-5}$ M, which proved to have an inhibitory activity on the radial growth of some phytopathogenic fungi, including *Fusarium oxysporum*, *Sclerotinia sclerotiorum*, and *Botrytis cinerea*, associated with an increase in hyphal branching. However, the concentrations that were found to be active were far higher compared with the "physiological" amounts of natural SLs produced and excreted by roots. More complete and exhaustive bioassays on a number of off-target organisms would be absolutely essential in respect of any SL practical applications (see above). To perform these bioassays, very large amounts of compounds would be necessary, and the high costs have probably made this kind of biotic evaluation not economically affordable yet. Beyond the scientific interest, Nijmegen-1 has been used in suicidal germination experiments in the field (Zwanenburg et al. 2016a). Currently, the high costs prevent the practical application of SL technology in the field. Interestingly, the search for new, efficient, and selective biological active compounds for field use can also be addressed by testing libraries of available compounds.

Glossary

Asymmetric induction A term applied to the selective synthesis of one diastereo-meric form of a compound resulting from the influence of an existing chiral centre adjacent to the developing asymmetric carbon atom.

Chiral auxiliary Is a stereogenic group or unit that is temporarily incorporated into an organic compound in order to control the stereochemical outcome of the synthesis. The **chirality** present in the **auxiliary** can bias the stereoselectivity of one or more subsequent reactions.

Chirality A term which may be applied to any asymmetric object or molecule. The property of nonidentity of an object with its mirror image.

Chromatography A series of related techniques for the separation of a mixture of compounds by their distribution between two phases. In gas-liquid chromatog-raphy, the distribution is between a gaseous and a liquid phase. In column chromatography, the distribution is between a liquid and a solid phase.

Circular dichroism The property (as of an optically active medium) of unequal absorption of right and left plane-polarized light so that the emergent light is elliptically polarized.

Configuration The order and relative spatial arrangement of the atoms in a mole-cule. Absolute configuration is when the relative three-dimensional arrangements in space of atoms in a chiral molecule have been correlated with an absolute standard.

Enantiomers A pair of isomers which are related as mirror images of one another.

Enantioselective synthesis, also called asymmetric synthesis A chemical reaction (or reaction sequence) in which one or more new elements of chirality are formed in a substrate molecule and which produces the stereoisomeric (enantiomeric or diastereoisomeric) products in unequal amounts.

Diastereomers (or diastereoisomers) Stereoisomeric structures which are not enantiomers (mirror images) of one another. Often applied to systems which differ only in the configuration at one carbon atom, e.g. *meso-* and *d-* or *l-*tartaric acids are diastereoisomeric.

Dextrorotatory The phenomenon in which plane-polarized light is turned in a clockwise direction.

Isomers Compounds having the same atomic composition (constitution) but differing in their chemical structure. They include structural isomers (chain or positional), tautomeric isomers, and stereoisomers—including geometrical isomers, optical isomers, and conformational isomers.

Mass spectrometry A form of spectrometry in which, generally, high-energy electrons are bombarded onto a sample and this generates charged fragments of the parent substance; these ions are then focused by electrostatic and magnetic fields to give a spectrum of the charged fragments.

Nuclear magnetic resonance (NMR) spectroscopy A form of spectroscopy which depends on the absorption and emission of energy arising from changes in the spin states of the nucleus of an atom. For aggregates of atoms, as in molecules, minor variations in these energy changes are caused by the local chemical environment. The energy changes used are in the radiofrequency range of the electromagnetic spectrum and depend upon the magnitude of an applied magnetic field.

Racemic mixture, racemate An equimolar mixture of the two enantiomeric isomers of a compound. As a consequence of the equal numbers of levo- and dextrorotatory molecules present in a racemate, there is no net rotation of plane-polarized light.

Resolution The separation of a racemate into its two enantiomers by means of some chiral agency.

Resonance The representation of a compound by two or more canonical structures in which the valence electrons are rearranged to give structures of similar probability. The actual structure is considered to be a hybrid or the resonance forms.

R,S convention A formal non-ambiguous, nomenclature system for the assignment of absolute configuration of structure to chiral atoms, using the Cahn, Ingold, and Prelog priority rules.

Stereochemistry The study of the spatial arrangements of atoms in molecules and complexes.

Stereoisomer Another name for configurational isomer.

Stereospecific reactions Reactions in which the stereochemistry of reagents affects the stereochemistry of products. Different stereoisomers as reagents give different stereoisomer as products.

References

Abe S, Sado A, Tanaka K, Kisugi T, Asami K, Ota S, Kim HI, Yoneyama K, Xie X, Ohnishi T, Seto Y, Yamaguchi S, Akiyama K, Yoneyama K, Nomura T (2014) Proc Natl Acad Sci 111:18084–18089

Akiyama K, Matsuzaki K, Hayashi H (2005) Nature 435:824–827

Berlage U, Schmidt J, Milkova Z, Welzel P (1987) Tetrahedron Lett 28:3095–3098
Bhattacharya C, Bonfante P, Deagostino A, Kapulnik Y, Larini P, Occhiato EG, Prandi C, Venturello P (2009) Org Biomol Chem 7:3413–3420
Boyer F-D, Germain ADS, Pillot J-P, Pouvreau J-B, Chen VX, Ramos S, Stevenin A, Simier P, Delavault P, Beau J-M, Rameau C (2012) Plant Physiol 159:1524–1544
Boyer FD, de Saint Germain A, Pouvreau JB, Clave G, Pillot JP, Roux A, Rasmussen A, Depuydt S, Lauressergues D, Frei Dit Frey N, Heugebaert TS, Stevens CV, Geelen D, Goormachtig S, Rameau C (2014) Mol Plant 7:675–690
Bromhead LJ, McErlean CSP (2017) Eur J Org Chem:5712–5723
Bromhead LJ, Smith J, McErlean CSP (2015) Aust J Chem 68:1221–1227
Brooks DW, Bevinakatti HS, Powell DR (1985) J Org Chem 50:3779–3781
Ćavar S, Zwanenburg B, Tarkowski P (2015) Phytochem Rev 14:691–711
Chen VX, Boyer F-D, Rameau C, Retailleau P, Vors J-P, Beau J-M (2010) Chem Eur J 16:13941–13945
Coggan P, Luhan PA, McPhail AT (1973) J Chem Soc Perk Trans 2:465–469
Cook CE, Whichard LP, Turner B, Wall ME (1966) Science 154:1189–1190
Cook CE, Whichard LP, Wall ME, Egley GH, Coggon P, Luhan PA, McPhail AT (1972) J Am Chem Soc 94:6198–6199
Dailey OD (1987) J Org Chem 52:1984–1989
Dieckmann MC, Dakas P-Y, De Mesmaeker A (2018) J Org Chem 83:125–135
Dvorakova M, Soudek P, Vanek T (2017) J Nat Prod 80:1318–1327
Flematti GR, Scaffidi A, Waters MT, Smith SM (2016) Planta 243:1361–1373
Frischmuth K, Samson E, Kranz A, Welzel P, Meuer H, Sheldrick WS (1991) Tetrahedron 47:9793–9806
Frischmuth K, Wagner U, Samson E, Weigelt D, Koll P, Meuer H, Sheldrick WS, Welzel P (1993) Tetrahedron Asymmetry 4:351–360
Fukui K, Yamagami D, Ito S, Asami T (2017) Front Plant Sci 8:936
Hauck C, Schildknecht H (1990) J Plant Physiol 136:126–128
Hauck C, Muller S, Schildknecht H (1992) J Plant Physiol 139:474–478
Heather J, Mittal R, Sih CJ (1974) J Am Chem Soc 96:1976–1977
Heather JB, Mittal RSD, Sih CJ (1976) J Am Chem Soc 98:3661–3669
Hirayama K, Mori K (1999) Eur J Org Chem 1999:2211–2217
Kadas I, Arvai G, Miklo K, Horvath G, Toke L, Toth G, Szollosy A, Bihari M (1996) J Environ Sci Health, Part B B31:561–566
Kim HI, Kisugi T, Khetkam P, Xie X, Yoneyema K, Uchida K, Yokota T, Nomura T, McErlean CSP, Yoneyama K (2014) Phytochemistry 103:85–88
Kisugi T, Xie X, Kim HI, Yoneyama K, Sado A, Akiyama K, Hayashi H, Uchida K, Yokota T, Nomura T, Yoneyama K (2013) Phytochemistry 87:60–64
Kitahara S, Tashiro T, Sugimoto Y, Sasaki M, Takikawa H (2011) Tetrahedron Lett 52:724–726
Kondo Y, Tadokoro E, Matsuura M, Iwasaki K, Sugimoto Y, Miyake H, Takikawa H, Sasaki M (2007) Biosci Biotechnol Biochem 71:2781–2786
Kumagai H, Fujiwara M, Kuse M, Takikawa H (2015) Biosci Biotechnol Biochem 79:1240–1245
Lace B, Prandi C (2016) Mol Plant 9:1099–1118
Lachia M, Dakas PY, De Mesmaeker A (2014) Tetrahedron Lett 55:6577–6581
Lachia M, Wolf HC, Jung PJ, Screpanti C, De Mesmaeker A (2015) Bioorg Med Chem Lett 25:2184–2188
Lombardi C, Artuso E, Grandi E, Lolli M, Spirakys F, Priola E, Prandi C (2017) Org Biomol Chem 15:8218–8231
Macalpine GA, Raphael RA, Shaw A, Taylor AW, Wild HJ (1976) J Chem Soc Perkin Trans 1:410–416
Matsui J, Yokota T, Bando M, Takeuchi Y, Mori K (1999a) Eur J Org Chem 1999:2201–2210
Matsui J, Bando M, Kido M, Takeuchi Y, Mori K (1999b) Eur J Org Chem 1999:2195–2199
Matsui J, Bando M, Kido M, Takeuchi Y, Mori K (1999c) Eur J Org Chem:2183–2194

Matsuura H, Ohashi K, Sasako H, Tagawa N, Takano Y, Ioka Y, Nabeta K, Yoshihara T (2008) Plant Growth Regul 54:31–36

Matusova R, Rani K, Verstappen FWA, Franssen MCR, Beale MH, Bouwmeester HJ (2005) Plant Physiol 139:920–934

Mori K, Matsui J (1997) Tetrahedron Lett 38:7891–7892

Mori K, Matsui J, Bando M, Kido M, Takeuchi Y (1997) Tetrahedron Lett 38:2507–2510

Mori K, Matsui J, Bando M, Kido M, Takeuchi Y (1998) Tetrahedron Lett 39:6023–6026

Morris JC, McErlean CSP (2016) Org Biomol Chem 14:1236–1238

Mwakaboko AS, Zwanenburg B (2016) Eur J Org Chem 2016:3495–3499

Nakamura H, Asami T (2014) Front Plant Sci 5:623

Oancea F, Georgescu E, Matusova R, Georgescu F, Nicolescu A, Raut I, Jecu ML, Vladulescu MC, Vladulescu L, Deleanu C (2017) Molecules 22. https://doi.org/10.3390/molecules22060961

Prandi C, Ghigo G, Occhiato EG, Scarpi D, Begliomini S, Lace B, Alberto G, Artuso E, Blangetti M (2014) Org Biomol Chem 12:2960–2968

Reizelman A, Scheren M, Nefkens GHL, Zwanenburg B (2000) Synthesis-Stuttgart 13:1944–1951

Samson E, Frischmuth K, Berlage U, Heinz U, Hobert K, Welzel P (1991) Tetrahedron 47:1411–1416

Scaffidi A, Waters MT, Sun YK, Skelton BW, Dixon KW, Ghisalberti EL, Flematti GR, Smith SM (2014) Plant Physiol 165:1221–1232

Seto Y, Sado A, Asami K, Hanada A, Umehara M, Akiyama K, Yamaguchi S (2014) Proc Natl Acad Sci 111:1640–1645

Sugimoto Y, Wigchert SCM, Thuring J, Zwanenburg B (1997) Tetrahedron Lett 38:2321–2324

Sugimoto Y, Wigchert SCM, Thuring J, Zwanenburg B (1998) J Org Chem 63:1259–1267

Takahashi A, Ogura Y, Enomoto M, Kuwahara S (2016) Tetrahedron 72:6634–6639

Takikawa H, Jikumaru S, Sugimoto Y, Xie XN, Yoneyama K, Sasaki M (2009) Tetrahedron Lett 50:4549–4551

Thuring JWJF, Nefkens GHL, Wegman MA, Klunder AJH, Zwanenburg B (1996) J Org Chem 61:6931–6935

Thuring J, Nefkens GHL, Zwanenburg B (1997) J Agric Food Chem 45:1409–1414

Ueno K, Nomura S, Muranaka S, Mizutani M, Takikawa H, Sugimoto Y (2011) J Agric Food Chem 59:10485–10490

Welzel P, Rohrig S, Milkova Z (1999) Chem Commun:2017–2022

Wigchert SCM, Kuiper E, Boelhouwer GJ, Nefkens GHL, Verkleij JAC, Zwanenburg B (1999) J Agric Food Chem 47:1705–1710

Xie X, Kusumoto D, Takeuchi Y, Yoneyama K, Yamada Y, Yoneyama K (2007) J Agric Food Chem 55:8067–8072

Xie X, Yoneyama K, Kusumoto D, Yamada Y, Takeuchi Y, Sugimoto Y, Yoneyama K (2008a) Tetrahedron Lett 49:2066–2068

Xie X, Yoneyama K, Kusumoto D, Yamada Y, Yokota T, Takeuchi Y, Yoneyama K (2008b) Phytochemistry 69:427–431

Yasui M, Ota R, Tsukano C, Takemoto Y (2017) Nat Commun 8:674

Yokota T, Sakai H, Okuno K, Yoneyama K, Takeuchi Y (1998) Phytochemistry 49:1967–1973

Zwanenburg B, Pospisil T (2013) Mol Plant 6:38–62

Zwanenburg B, Nayak SK, Charnikhova TV, Bouwmeester HJ (2013) Bioorg Med Chem Lett 23:5182–5186

Zwanenburg B, Mwakaboko AS, Kannan C (2016a) Pest Manag Sci 72:2016–2025

Zwanenburg B, Regeling H, Van Tilburg-Joukema CW, Van Oss B, Molenveld P, De Gelder R, Tinnemans P (2016b) Eur J Org Chem 2016:2163–2169

Zwanenburg B, Pospíšil T, Ćavar Zeljković S (2016c) Planta 243:1311–1326

Printed in the United States
By Bookmasters